THE COMPLETE BOOK OF
HOLOGRAMS

How They Work and How to Make Them

Joseph E. Kasper
and
Steven A. Feller

Wiley Science Editions

John Wiley & Sons, Inc.

New York • Chichester • Brisbane • Toronto • Singapore

This book was originally published under the title *The Hologram Book* in 1985, by Prentice-Hall, Incorporated, Englewood Cliffs, NJ.

Figure 2-1 and quotations from *Treatise on Light* by Christiaan Huygens are used with permission of the publisher, Dover Publications (New York, 1962).

Quotations in the Preface and in Chapter 11 from the Nobel Prize Address of Dennis Gabor are used with permission of the Nobel Foundation. The address is copyright 1972 by the Nobel Foundation.

Quotations in Chapter 7 from "Geometric Model for Holography" by Tung H. Jeong, *American Journal of Physics*, Vol. 43, No. 8, August 1975 are used with the permission of the copyright owner, The American Association of Physics Teachers, copyright 1975, and Dr. Jeong, who further kindly granted permission to use all information he has published.

The quotation in Chapter 10 from "Demonstration Holograph for Comparing an Image Lens and a Real Lens" by Mac Rugheimer and Larry D. Kirkpatrick, *American Journal of Physics*, November 1977 is used by permission of the copyright owner (The American Association of Physics Teachers) and Mac Rugheimer and Larry D. Kirkpatrick.

Figures 1.1, 5.1, and 10.3 are used with the permission of Mr. Rainbows, P.O. Box 27056, Philadelphia, PA 19118.

Figure 11.1 was prepared from a special copy of an original holographic micrograph of an argon atom, which was most kindly loaned by Dr. L.S. Bartell, Department of Chemistry, University of Michigan, Ann Arbor, Michigan, with his permission to use it.

Library of Congress Cataloging in Publication Data:

Kasper, Joseph Emil, 1920-
    The complete book of holograms.

    (Wiley science editions)
    1. Holography.   I. Feller, Steven A.   II. Title.
III. Series.
TA1540.K37   1987   621.36'75   87-16209
ISBN 0-471-62941-3

Printed in the United States of America
87   88   10   9   8   7   6   5   4   3   2   1

Contents

Foreword

One of the pleasures of my professional life has been knowing the senior author of this book, Joseph E. Kasper, first when he was a graduate student at the University of Iowa and then for many years as a member of the faculty of nearby Coe College. *The Hologram Book* exemplifies the quality of that pleasure. To paraphrase a remark that I made at a public occasion a few years ago: "When Kasper and Feller explain a physical phenomenon, they *really* explain it . . . with depth, sophistication, and clarity."

A hologram is a photographic record of the interference pattern of two superimposed beams of coherent light, one directly from the source and the other reflected or scattered from a physical object. The proper viewing of a hologram is an eerie and mystical experience. No curious person can fail to ask for an explanation. It is for such a person that this book has been written.

The authors treat the subject with rigor and loving care but with minimal mathematics. Every sentence is turned out with expository pride. The reader is treated as an uninformed but interested and intelligent friend. There is no hint of condescension. There are no it-is-obvious-thats or it-can-be-shown-thats. Explanations are lucid and full. The exposition proceeds chapter by chapter in easy stages, replete with clear diagrams that are closely integrated with the text. A casual reader might be satisfied with the first eight chapters, but the book is like a good mystery story and a truly interested reader, as I was, can scarcely wait to find out what happens next. The latter portion of the book descibes the multifold applications of holography in modern technology, and last, but not least, there are two detailed and excellent chapters for the experimentally inclined person on "How to Make Holograms."

I predict that Kasper and Feller will become a standard reference on holography for students and interested laymen. It might well be called "The Compleat Book of Holograms."

JAMES A. VAN ALLEN
Carver Professor of Physics
University of Iowa

Preface

In his acceptance speech when awarded the Nobel prize for his invention of holography, Dr. Dennis Gabor said, "I have the advantage in this lecture . . . that I need not write down a single equation or show an abstract graph. One can, of course, introduce almost any amount of mathematics into holography, but the essentials can be explained and understood from physical arguments." Indeed, Dr. Gabor could have said, ". . . understood from *simple* physical arguments."

It is our goal to serve readers with no background in the science of optics with a systematic but easily read book on holography that explains those physical arguments. We believe that there are many people such as students, artists, photographers, hobbyists, and others who want a satisfying explanation of holography. Their question "How do holograms work?" is one to which this book is addressed.

The book was written to be usable in an introductory course at the high school or college level and also to be pleasantly readable by any interested reader on his or her own. According to an ancient saying, there is no royal road to geometry, and equally there is none to holography, but anyone can come to a sound basic understanding of the subject by proceeding step by step down what is a fairly short road.

A special feature of this book is its use of what is called the "geometric model" of holography originated by Dr. T. H. Jeong. This is a simple and intuitively appealing model. Because many properties of holograms depend on what are called diffraction phenomena, with which the geometric model does not deal, we also explain the older "zone plate" model.

We have included extensive discussions of applications of holography, some of which are in common practice today, and some, such as holographic movies and television, that are still undergoing development.

We have also had in mind readers who may want to try their hands at making holograms. For those readers, we have included descriptions of a number of different ways to make holograms and practical advice about equipping a holographic lab. In doing this, we have stressed the particular method that we have found to be simple, effective, and relatively inexpensive. For those who wish to go further in experimentation, we offer additional assistance.

May readers enjoy this book as we have enjoyed writing it!

This book is dedicated to the students of holography who have used this book while it was being prepared and helped greatly to improve it—and to all persons who will use it in the future.

Holography and ordinary photography

A hologram is a photographic emulsion in which information about a scene is recorded in a very special way. When the hologram is properly illuminated, you, the viewer, see a realistic, three-dimensional representation of the scene. Upon seeing your first hologram, you are certain to feel puzzlement and disbelief. You may place your hand where the scene apparently lies, only to find that nothing tangible is there. You may then look around, trying to learn what the trick or gimmick is. Holograms are indeed provocative; you will not lose the sense of wonder even after having seen many of them.

The trick in holography is in the special manner optical information about a scene is recorded in the emulsion. This is done in such a way that the information is *complete,* so that the scene can be made visible in all its true spatial 3-D aspects with shadows and varying intensities through the scene reproduced realistically. In fact, the word *holography* is derived from Greek roots meaning "complete writing."

Holograms are not arcane things, shrouded in mystery. On the contrary, they are understandable devices operating on the basis of simple optical principles. This book explains those principles and shows how they account for the properties of holograms. This first chapter discusses some general aspects of holography that should be known before you begin to study the underlying optics in Chapter 2.

1

The Holographic Scene

One way to view a holographic scene is shown in Figure 1.1. The hologram itself is a sheet of exposed and developed photographic emulsion. The light from a laser in the background shines onto the plate. While on the near side of the plate you look *through* the plate and perceive objects in the scene very much as though you were looking through a window at an actual scene lying beyond the window.

A flat picture such as Figure 1.1 cannot show how striking the real view is. It is immediately obvious that you are seeing something very special. The objects in the scene have *depth*—a three-dimensional appearance. As you move your head, you can see around objects and can see things that from other points of view lie behind other objects. Shadows change their positions. Light seems to bounce off objects at varying angles. A diamond ring in a holographic

Figure 1.1. *The virtual image produced by a transmission hologram, illuminated from behind by a laser. Of course, this photograph cannot convey the vivid three-dimensional nature of the actual scene. (Photo by Dr. C. E. Hamilton, produced by Mr. Rainbows.)*

scene can sparkle and show changing aspects of its facets. (One of the first holograms made by the wife of one of the authors was of her engagement ring. It was a striking hologram.)

The particular kind of hologram in Figure 1.1 is called a *transmission hologram* because the laser light is transmitted through the hologram to the viewer's eyes. Another principal kind, the *reflection hologram,* is viewed in light reflected from a light source on the viewer's side into his or her eyes.

The Hologram Itself

The hologram is a layer of photographic emulsion, either on clear flexible plastic material or on a glass backing.[1] For simplicity we will refer to the hologram as a *plate.* Inspection of the plate reveals a remarkable fact. There is nothing on it that resembles the scene. The plate may be quite clear or somewhat cloudy, or it may have dark swirls and striations.

When you look at a hologram such as Figure 1.2 the obvious question is, "Where is the scene?" In fact, the scene is not there at all; rather, it is

Figure 1.2. *The appearance of a hologram plate (unbleached). The swirls and striations seen are not the microscopic interference patterns that really constitute the hologram. They are due to dust or imperfections in the optical apparatus that was used to make the hologram. (Photo by Dr. C. E. Hamilton.)*

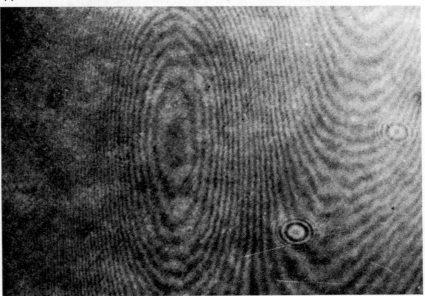

_____
[1]In recent times substances other than the familiar silver halide photographic emulsions have been developed for making holograms.

information about the scene, coded in the form of *interference patterns,* that is recorded in the hologram. The coded information itself would have no discernible resemblance to the scene even if you could see it with the naked eye. Actually, the interference patterns are present on a microscopic scale. If you happen to see swirls or other markings as in Figure 1.2, these are caused by extraneous effects such as dust particles or imperfections in some optical apparatus used in making the hologram.

It is only when the hologram is suitably illuminated that the information contained in the hologram can be decoded and the scene *reconstructed* or made visible.

MAKING PHOTOGRAPHS AND HOLOGRAMS

It must be a rare person who does not know something about the making of ordinary photographs. Nevertheless, we outline the process here. In Figure 1.3 a camera is used to photograph an object (a simple cylinder). There is a source of light, and some of the light falls on the object and reflects into the lens of the camera. The lens focuses the light onto photographic emulsion so that a sharp image will result. A shutter (not shown) controls the length of time of exposure of the emulsion. After exposure, the film is removed from the camera and is processed. The immediate result is a negative from which a positive print can be made.

Figures 1.4 and 1.5 show one arrangement for making a hologram. Let us first describe the configuration in general and then in more detail.

As in ordinary photography, there must be a source of light to illuminate the scene and activate the photographic emulsion. The light source shown is

Figure 1.3. Making an ordinary photograph.

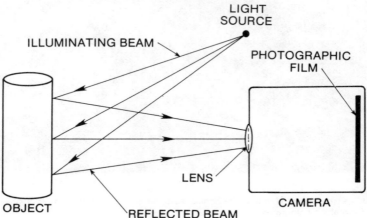

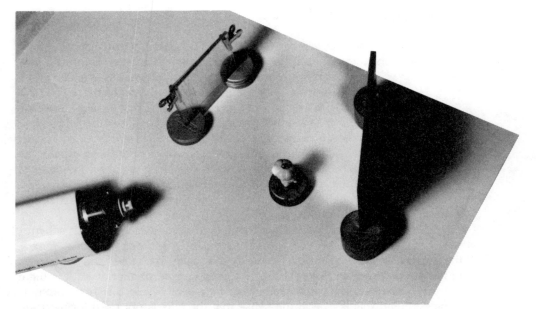

Figure 1.4. *Set-up of apparatus on a holographic table top for making a one-beam transmission hologram. Laser at lower left; front surface mirror at right; film holder containing film at upper left; object (a small doll) in center. In this compact arrangement a strongly diverging lens is used at the port of the laser. More commonly the laser is placed farther off to the left. (Photo by Dr. C. E. Hamilton.)*

Figure 1.5. *One way to make a hologram.*

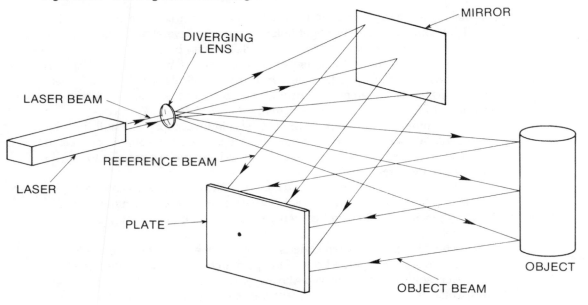

a laser, for reasons that will be made clear. The laser beam is narrow, only a few millimeters across, and must be spread out so that it can illuminate substantial areas. The divergence of the beam is accomplished by a lens.

Now something of great importance is done. Some of the spread-out beam of laser light falls on a flat mirror that is oriented so that the light reflecting from it hits the plate, while another part of the beam bypasses the mirror, falls on the object, and reflects back to the plate. These are the *reference* and *object beams,* respectively. When the shutter (not shown) is opened to expose the plate, these two beams of laser light interact with each other. An interference pattern is formed and is recorded in the emulsion. After the plate is processed, you have a permanent record in the form of a negative of the interference pattern. This is a hologram.

Before proceeding to the fundamentally important aspects of this scheme, two simpler matters deserve comment. In making a photograph with a camera, the lens is essential in getting an image of the scene. In contrast, the lens shown in the arrangement in Figure 1.5 is not used in that manner at all. It is there only to spread out the laser beam. You may hear it said that holography is "lensless," and this is true in the sense that the lens is not used for focusing the image. Second, special film must be used in holography because the interference pattern to be recorded in it is on a microscopic level of fineness. Photographers would call this *very high resolution film.*

COHERENCE AND THE LASER

In making a photograph, as in Figure 1.3, there is nothing special required of the light source. An ordinary light bulb will serve, or several bulbs, as might be used in an interior room. Sunlight or candlelight can be used.

In contrast, the holographic method *always* requires that two special conditions be met. One is that there must be specifically *two* beams incident on the plate—the reference beam and the object beam. The two beams are required to form the interference pattern.

The other special property of light used in holography is that it must be *coherent.* We will have more to say about this later, but you will be helped considerably by a brief explanation now. This will also make it possible to explain the role of the laser in holography.

Let us begin by saying that you cannot make a hologram with a flashlight, the sun, or any ordinary source of light. What we call "ordinary" sources produce light that lacks coordination between parts of the light beam. They are said to be *incoherent* sources.

The difference between coherence and incoherence can be explained by a simple analogy. Imagine a body of troops marching in orderly ranks and files, with every soldier perfectly in step with every other soldier. In optical terminology this would be considered a *coherent* formation. If, on the other hand, there were no synchronization between one soldier's pacing and that of another, the formation would be *incoherent.* Now consider a beam of light. The light consists of electromagnetic waves. Corresponding to a file of marching soldiers there is an oscillation in the beam along the direction of travel, and corresponding to the ranks of soldiers there are oscillations across the breadth of the beam. This is illustrated in Figure 1.6. If all the oscillations are perfectly regular and all are in step with all the others, the light is coherent. In the case of incoherent light, such synchronization among parts of the light beam is not present.

Why is it necessary to use coherent light in making holograms? Suppose that two beams of coherent light are superimposed, as are the reference beam and the object beam in Figure 1.5. They interact with each other in the plate, producing a very finely detailed pattern of exposed and unexposed regions, and this is the hologram. If at some point the two light beams are completely in step with each other throughout the exposure, there will be a maximum blackening. If at some point the two beams are completely out of step with each other during the exposure, they will cancel out each other's effect and there will be no darkening. In this way, alternating light and dark regions (interference fringes) are formed. This is the desired result, but it would be spoiled if the requirement of coherence were not met. Suppose that at some point the light beams were sometimes in step and sometimes out of step, with random fluctuations in the relationship. If this situation extended over all of the emulsion, the processed plate would show only a darkening, and no hologram would be produced.

Where can you get the coherent light required? At the present time, there is only one satisfactory source: the *laser.* A laser is a device that produces a

Figure 1.6. *Schematic representation of a beam of fully coherent light. The arrows (a, b, c, d) indcate direction of travel. Along each "file" the waveform has its peaks and valleys uniformly spaced. Along each "rank" (as along A and B) the waves are in step.*

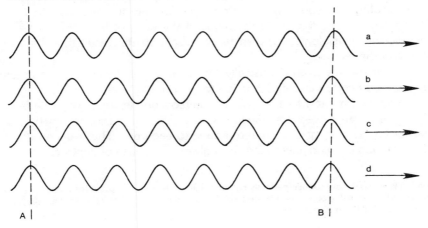

highly coherent beam of light.[2] Fortunately, the common helium–neon (He-Ne) gas laser can now be obtained with sufficient power at a cost as low.as about $300—and it is simple to operate, long-lived, and safe. Most holograms today are made with the He-Ne laser.

A laser beam has two other special characteristics. One of these is that the light is highly *monochromatic,* which means that the color is very sharply defined. In more technical terms, the light is very nearly of one single wavelength. For the He-Ne gas laser the color is a red that resembles that used in neon advertising signs. It happens that the coherence and the monochromaticity of the laser beam are related. This will be explained later.

The other special property of a laser beam is that it is highly collimated. It issues from the laser with a small cross-sectional dimension, usually a few millimeters. As the beam travels along, this dimension increases very little, even over very great distances. Although the coherence and purity of color of the laser light are very important in holography, the collimation is an obstacle. As we have seen, a setup for making a hologram begins by spreading the beam out with a lens.

There are those who think every laser beam is extremely intense. Actually, the beam may be only as intense as that from a flashlight, intense enough to vaporize metal, or anything in between those extremes. The only importance of the intensity of the laser beam in holography is that it be great enough to permit exposure time of the photographic emulsion to be reasonably short, and low enough to do no damage to the objects (which might be people!) being holographed.[3]

VIEWING PHOTOGRAPHS AND HOLOGRAMS

Figure 1.7 represents an eye looking at an ordinary photograph. Light from an ordinary source such as the sun or a room lamp casts light onto the surface of the photograph, some of the light reflects off into the eye, and the lens in the eye focuses the light onto the retina at the back of the eye. Then a very complicated physiological system takes over, with the result that the viewer perceives the picture.

What you see is an intrinsically flat surface. Only by complicated mental processes do you relate what you see to the original scene.

You are probably familiar with the stereoscope in which two side-by-side photographs are viewed simultaneously, or with 3-D television in which colored glasses are used to give a different image to each eye. These depend on the phenomenon of "binocular vision" and give some sense of depth in the scene,

_____

[2]Even a laser does not produce perfectly coherent light. There can be excellent coherence across the beam but inevitably only over a limited extent along the beam. Chapter 3 explains this further.
[3]See "Laser Safety" in Appendix I.

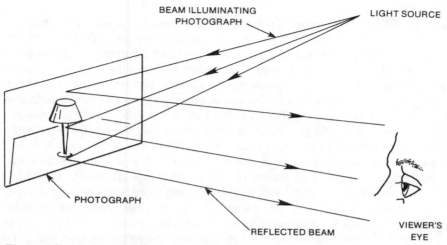

BEAM ILLUMINATING PHOTOGRAPH

LIGHT SOURCE

PHOTOGRAPH

REFLECTED BEAM

VIEWER'S EYE

Figure 1.7. *Viewing an ordinary photograph.*

but they do not meet the acid test for three-dimensional realism: Can you shift your point of view and look around a tree, for example, and see objects that are behind it?

Now let us consider Figure 1.8 which is a diagrammatic version of Figure 1.1. It shows how to view the scene with a transmission hologram.[4] The laser beam is diverged by a lens to spread it enough to illuminate the plate. Your eye picks up the light on the other side of the hologram, after it has been acted upon or "decoded." You *can* look around the tree.

A hologram differs from a photograph in a number of striking ways other

Figure 1.8. *Reconstructing the scene holographically.*

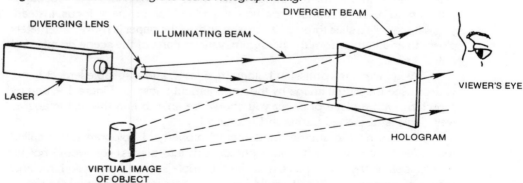

DIVERGING LENS

DIVERGENT BEAM

ILLUMINATING BEAM

VIEWER'S EYE

LASER

HOLOGRAM

VIRTUAL IMAGE OF OBJECT

---

[4]If the hologram were of the *reflection* type, the arrangement for viewing would resemble that in Figure 1.8, but the light source would be on the viewer's side, as it would be if he or she were looking at a mirror.

than the realism of the image it provides. One of these is that while a photograph is a single representation of a scene, a hologram gives more than one image. Figure 1.9 provides a basis for this discussion. The hologram is a transmission hologram illuminated by a diverged laser beam. Some of the laser light passing through the plate is acted upon by the interference pattern recorded so that it emerges as a diverging beam. This is suggested by the rays labeled A, B, and C. When your eye receives this light, the lens in the eye is used to focus the light on the retina, and you interpret what you see as though the light came from a nonexistent *virtual* source. The eye reacts as though the rays A, B, and C were extensions of the fictitious rays A′, B′, and C′. These fictitious rays would exist if there were a real object where the "virtual image" is shown. The object is "seen" as if it were at that location, even though there is no real object there.

Actually, looking at a virtual image is an everyday experience for most people and evokes no mystery. When you look into a mirror, you see an image of yourself as though there were a person behind the mirror while in reality the mirror is opaque (though reflecting). Light from a source on your side of the mirror reflects into your eyes and you interpret what your eye receives as a "virtual image."

Some of the laser light comes from the hologram as a *converging* beam. This is suggested by the rays D, E, and F in the drawing. These come to a focus and produce a second image called the *real image*. This means that the light (rays D, E, F and others) converges, focuses, and creates an optical replica of the object. If you were to put a piece of paper there, you could look at the paper and see the image. No focusing lens is needed to help the rays produce this image, unlike the case of the virtual image. Since the image is really there, it can be viewed with the eye, as when looking at a photograph.

Like the virtual image, the real image is fully three-dimensional. A piece of paper put there to reveal it cannot show you all parts of the image in focus. Instead, you will see only a slice through the image and, as the paper is moved alternately closer to and farther from the hologram, various parts of the scene go into and out of focus.

If you have a transmission hologram, a laser, and a suitable lens at hand, you can see the virtual image by the arrangement shown in Figure 1.9, and by searching with a piece of paper you should be able to find the real image as well. (This is discussed further in Chapter 5.)

We turn now to another fundamental property of holograms. It is called *redundancy*. If you were to cut a photograph into pieces, you would not be able to look at any single piece and see the whole scene. You would have only pieces of a jigsaw puzzle that would have to be reassembled properly to reproduce the scene. On the other hand, if you were to break a hologram into pieces or, more practically, use a card with a hole in it to block off all but a small segment of the hologram, you would still get a realistic three-dimensional view of the scene from the segment. You can correctly think of viewing a holographic

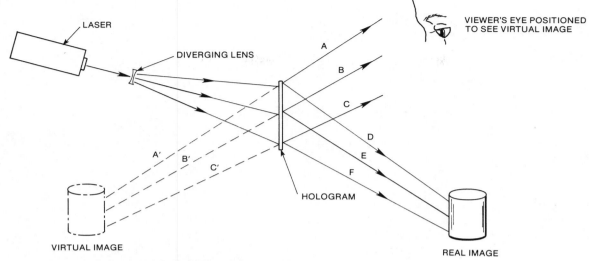

Figure 1.9. *The hologram produces two images.*

scene as like looking through a window. If you use only a small piece of a hologram, this is like looking through a small window, while using the whole hologram is like using a larger window. You would be limited in moving your head around to view the scene, but what you see would otherwise be fully realistic. Thus a hologram records information about the whole scene in each small piece of the plate.

Attempts to understand how the human brain stores information have heretofore been based on the assumption that it is a localized process, with each piece of information stored in a definite place, as in the memory of a computer. There are difficulties with this idea; removal of part of a brain does not necessarily remove a definite portion of memory. It has been suggested that the human memory is perhaps like a hologram in that what is stored is spread out over whole regions, with a small portion of the brain able to retain the information.

CONCLUDING REMARKS

We have tried to give you a general idea of what holograms are, how they are made, and how they are used to reconstruct the scene. There are other kinds of holograms, other techniques for making them, and other interesting properties inherent in them. However, these matters are best discussed after you learn more about wave theory, which begins in the next chapter.

11

The interference
of water waves

Holography is an application of the wave theory of light. In fact, it was first conceived as a theoretical possibility on that basis by the British scientist Dennis Gabor in 1947. His idea was to combine two sets of coherent beams of light, to record the interference pattern produced by their interaction in a photographic emulsion, and to do this in such a way that the plate would then contain such complete (or *holo-graphic*) information about a scene that with suitable illumination of the plate a realistic image of the scene could be reconstructed. Although Dr. Gabor had only comparatively unsatisfactory light sources with which to work (the laser became available much later in 1960), he was nevertheless able to make rudimentary holograms to establish the validity of his idea. For this pioneering work, Dr. Gabor was given the Nobel prize in physics in 1971.

Both this brief historical sketch and the discussion of holograms in Chapter 1 indicate the importance of the interference of light beams for understanding holograms. This chapter discusses some basic properties of waves and of interference in particular.

This chapter will deal only with water waves as the behavior of surface waves on a body of water is especially easy to understand. Chapter 3 deals with light waves. Chapter 4 applies these optical concepts to the main question: "How do holograms work?"

CHRISTIAAN HUYGENS, WATER WAVES, AND LIGHT WAVES

In the late 1600s—some 300 years ago—a Dutch scientist named Christiaan Huygens wrote a brief book called *Treatise on Light*.[1] In this book, Huygens proposed a view of the nature of light that is still fresh and productive in its insight. We can do no better than to let Huygens speak:

Huygens first makes it clear that he has a correct view of how *sound* is propagated through air.

> We know that by means of the air . . . sound spreads around the spot where it has been produced, by a movement which is passed on successively from one part of the air to another; and that the spreading of this movement, taking place equally rapidly on all sides, ought to form spreading spherical surfaces ever enlarging and which strike our ears.

At once he applies this idea to the propagation of *light* from a source to the eyes.

> Now there is no doubt at all that light also comes from the luminous body to our eyes by some movement impressed on the matter which is between the two. . . . If, in addition, light takes time for its passage . . . it will follow that this movement, impressed on the intervening matter, is successive; and consequently it spreads, as sound does, by spherical surfaces and waves; for I call them waves from their resemblance to those which are seen to be formed in water when a stone is thrown into it, and which present a successive spreading in circles, though these . . . are only in a flat surface.

Figure 2.1 is a reproduction of Huygens' own drawing which he drew to illustrate his ideas. He explains that "each little region of a luminous body, such as the Sun, a candle, or a burning coal, generates its own waves of which that region is the centre. Thus in the flame of a candle, having distinguished the points A, B, C, concentric circles described about each of these points represent the waves which come from them. And one must imagine the same about every point of the surface and of the part within the flame."

Huygens was incorrect in his construction in one respect. He knew that sound is propagated in a tangible medium—air—and that water waves are

---

[1]Christiaan Huygens, *Treatise on Light,* (New York: Dover Publications, Inc., 1962). Figure 2.1 and quotations from this work are used with permission of Dover Publications, Inc.

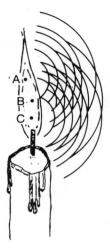

Figure 2.1. _Huygens' conception of light wavefronts. (From_ Treatise on Light, _by Christian Huygens, New York: Dover Publications Inc., 1962. Used with permission.)_

propagated on the surface of a tangible medium—water. He therefore conjectured that light must also be propagated through a tangible medium of some kind. Today we have a more sophisticated theory of light in which light consists of electromagnetic fields that travel in vacuum, rather than in a material medium. This theory was developed in the 1800s, long after the time of Huygens by the English physicist James Clerk Maxwell. This will not concern us in this book, for no use of the modern electromagnetic theory of-light is needed for an understanding of holography.

Huygens was influenced strongly by his familiarity with water waves. The word _waves_ tends to bring to mind images of ocean waves, rolling surf, or the like. There are profound differences between water waves and light waves, as already indicated, but it is remarkable to what an extent they are analogous.

WATER WAVES FROM A POINT SOURCE

If you dropped a stone into water, the results would look like the sketch in Figure 2.2. Circular ripples move outward from the point of disturbance. As the ripples move outward there is also an up-and-down movement of the surface, perpendicuar to the direction of travel. For this reason, such waves are said to be _transverse._ (In contrast, sound waves in air consist of regions of compression and rarefication along the direction of travel and are called _longitudinal_ waves.) That water waves are transverse is shown convincingly by considering a bobber on the surface, rising and falling vertically as the traveling waves pass it.

Suppose you want a way to produce more regular and controllable water waves than dropped stones can provide. Let's assume that a ripple tank is used, as in Figure 2.3. This is a tank of water with a small mechanical device mounted

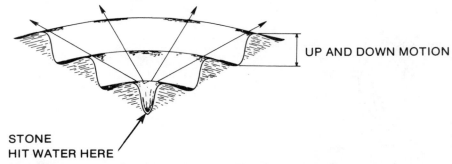

MOTION OF WAVES RADIALLY OUTWARD

UP AND DOWN MOTION

STONE
HIT WATER HERE

Figure 2.2. *Spreading motion of transverse waves on a water surface.*

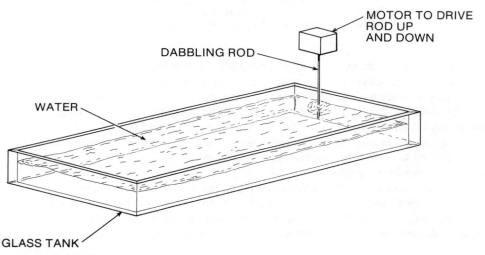

MOTOR TO DRIVE
ROD UP
AND DOWN

DABBLING ROD

WATER

GLASS TANK

Figure 2.3. *A ripple tank.*

above the surface. The device consists of a pointed rod that is driven up and down by a motor with great regularity. The result is a continuous "wave train" from what is appropriately called a *point source*. The tank is usually made of glass so that light can be shone upward or downward through it, helping to make the waves on the surface visible.

Figure 2.4 (a) is a representation of what you would see. Circular waves, each concentric on the point source, move radially outward. As the waves pass any one point, there is a transverse disturbance, or bobbing up and down. Along any solid circle shown in the sketch, there is a crest in the wave pattern. Between any two consecutive crests is a circle along which there is a valley in the wave

16

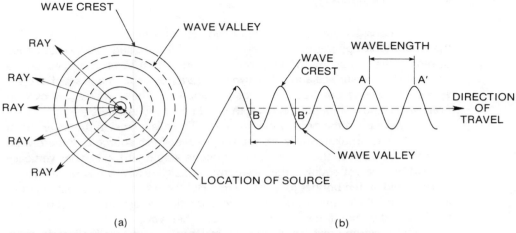

Figure 2.4. *(a) Overhead view of spreading wavefronts. (b) Side view.*

pattern; these are indicated by dotted circles. As a matter of terminology, let us call any of the circles a *wavefront.*

In order to indicate the radially outward direction of travel of the waves, arrows are used. These are called *rays.*

Let's assume that the ripple tank rod touches the surface with perfect regularity. In that case, the wavefronts will be uniformly spaced, and the distance between any two successive crests or any two successive valleys, or any two corresponding points, will be well-defined. This distance is called the *wavelength.* Part (b) of Figure 2.4 shows a side-on view of a wavetrain along a ray for further clarity. The wavelength is the distance A-A', or equally well, the distance B-B'.

Anywhere on the water surface there is a regular oscillation up and down as the waves pass by. The time rate at which this occurs is called the *frequency* of the waves. This is the same as the time rate at which the rod dabbles the water.

Finally, it is important to know what is meant by *displacement* and *amplitude* of a water wave. It is useful to think again of a bobber on the water, moving up and down as the waves pass by. At any instant the bobber may be at its normal undisturbed level, higher or lower. The vertical distance above or below the normal water surface level corresponds to the *displacement* of the wave. The displacement is a continuously varying quantity, increasing upward, decreasing to zero, increasing in magnitude downward, returning to zero, and repeating the cycle. What is meant by the *amplitude* of the wave is simply the maximum vertical displacement from the zero level, either upward or downward.

17

THE NATURE OF INTERFERENCE

Now let's consider the *interference* of two sets of water waves. In Figure 2.5, the ripple tank apparatus has been modified by installing two pointed rods driven by the motor. As the rods move up and down in unison, their points agitate the water surface and act as two *coherent* point sources. A set of circular waves will spread out from each of the sources. These waves will overlap everywhere and will interact in ways that will now be described.

Consider first the point *a* in Figure 2.6. There is a crest in the waves from source A, and simultaneously a crest in the waves from source B. If there were a bobber at point *a*, the A waves acting alone would lift the bobber vertically upward to a height equal to the wave amplitude. Similarly, the B waves acting alone would lift the bobber to a like height. But the A and B waves act together actually, and their effects *add*. Their collaboration would lift the bobber to a height equal to two times the amplitude of either wave.

This crest-on-crest condition is an instance of *constructive interference*. This kind of interference can occur at *any* point if the two superimposed waves are completely in step.

Let us elaborate on this general definition a bit more to help make its meaning clear. At point *b* there is simultaneously a valley in the A wave and in the B wave. The wave effects on the bobber act collaboratively—now adding together *downward*—and the bobber falls to a depth equal to two times the amplitude of either wave. Thus valley-on-valley is also an instance of constructive interference. It is easy to forget in looking at the figure that it is only a snapshot of oscillating and moving waves. Actually, as time passes, the bobber at point

Figure 2.5. *Ripple tank modified to provide two coherent point sources of water waves.*

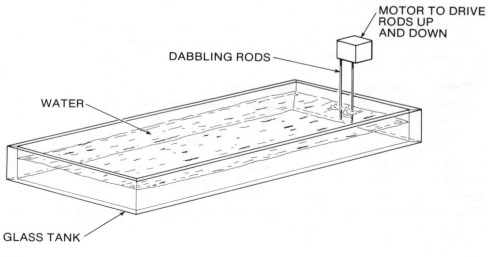

MOTOR TO DRIVE
RODS UP
AND DOWN

DABBLING RODS

WATER

GLASS TANK

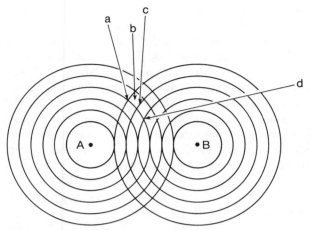

Figure 2.6. *Circular wavecrests from two coherent point sources.*

a undergoes up-and-down motion over and over. Thus, at some time later than the instant shown, there will be a valley-on-valley at point *a* and crest-on-crest at point *b.*

The opposite case is known as *destructive interference.* An illustrative instance occurs at point *c.* The B wave has a crest there and it alone would push a bobber upward to a height equal to one wave amplitude. At the same time, the A wave is at its valley and would strive to lower the bobber to a like depth. Since the forces on the bobber would be equal and oppositely directed, the bobber would have zero net displacement. It is important to realize that a bobber at *c* would *never* move up and down, because the passing A and B waves would at all times cancel each other out.

Finally, at intermediate points there would be an intermediate condition called *partial interference.* This happens at point *d,* for a representative case. The B wave is cresting there, but the A wave is not at its crest or its valley. The A wave would strive to push the bobber upward with a strength less than its maximum. At point *d* the A and B waves never collaborate fully as they do at *a* and at *b,* nor do they ever oppose each other fully as they do at *c.*

PERMANENCE IN THE MIDST OF COMPLICATED MOTIONS

Figure 2.6 represents the waves at one particular instant of time. In the real situation, conditions are highly dynamic. There are up-and-down movements of various parts of the water surface and there are general wave motions outward from the two point sources. In spite of this richness of motion over the surface, the interactions of the waves result in a very important effect—the overall pattern on the surface of the water remains stationary.

Consider once again points *a* and *c* in Figure 2.6. In the course of time, there would be an oscillation up and down of the water surface at *a* and this would reach a maximum extreme in both vertical directions. There would never be any oscillation vertically at *c*. There are many other points where there are maximal oscillations and many where there are never any oscillations.

At point *d* there is partial interference. The water surface oscillates up and down but never to a maximal extent as would occur if the interference were completely constructive. The conditions at *d* and at the many other points where the interference is only partial are also permanent.

THE INTERFERENCE PATTERN IN MORE DETAIL

In Figure 2.7, A and B are point sources of coherent water waves, as before. The spreading waves overlap and interfere, constructively in some places and destructively in others, and in intermediate fashion at in-between points. Our concern now will be with the overall stationary interference pattern that results.

Let's begin with the line labeled L. At all points along this line, there is constructive interference. Physically this means that all along this line the water surface oscillates up and down with maximum amplitude. The reason for this is that each point on the line is equally distant from each of the sources A and B, and so the A and B waves, traveling as they do at the same speed, must always arrive at the point in step with each other.

The curve labeled C is also a "locus" of points of constructive interference. At any point on it the waves from the two sources are in phase at all times. A few points are shown at which the crest-on-crest condition is met. The curve C′ is another such locus. It is the counterpart of the curve C.

Figure 2.7. *Some details of the stationary interference pattern.*

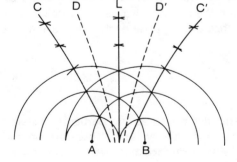

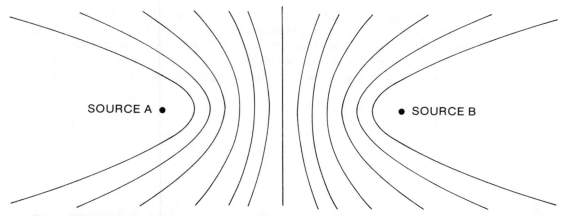

Figure 2.8. *The appearance of the interference pattern. The curves map out the regions of maximum amplitude of the water waves.*

On the other hand, the curve labeled D and its counterpart D′ are such that there is destructive interference at all points on each, at all times. It should be easy to identify points along either where there is the crest-on-valley condition, for example.

After these considerations, let's turn to Figure 2.8. This drawing represents the *main* features of what you would see if you looked at the water surface during an actual ripple tank experiment. The curves map out the regions of maximum amplitude in the interference pattern. They reflect the light that is being shone on the water differently from the other regions, so you can discern them. In this drawing the wavelength and the distance between the sources A and B are related differently from Figure 2.7, so there are more interference fringes between the sources in Figure 2.8 than in Figure 2.7, but the same fundamental physical ideas are used in understanding both cases. (In a real experiment, the fringes would be seen as somewhat broad. Also there would be inevitable unwanted reflections from the walls of the tank and possibly other disturbances. For clarity the drawing has been simplified.)

REMARKS ABOUT COHERENCE AND INCOHERENCE

Interference effects with water waves can be seen not only with a ripple tank or some other highly controlled arrangement. They can be seen when random raindrops or perhaps the movements of small insects produce spreading waves that overlap. However, in such cases the interference pattern is not constant. The fringes change in time—moving, fading away here and forming there. Raindrops and insects are *incoherent* sources. In the ripple tank experiment it is the continuous regularity of the successive disturbances of the surface by each rod, and the perfect synchronization between the actions of the two rods which are necessary for the production of a stationary interference pattern. The first of these conditions is summed up by saying each rod is a *coherent* source and the second by saying that the rods are *mutually coherent.*

The interference
of light waves

This chapter capitalizes on what was developed in Chapter 2, where only the behavior of water waves was considered. Now we extend the concepts to make them apply to light waves.

WAVELENGTH, FREQUENCY, AND WAVE VELOCITY

Wavelength and frequency are interpreted for light waves just as they were for water waves. The wavelength of a light wave is the length of any one complete oscillation; the frequency is the time rate at which the oscillations occur.

The numerical values of wavelengths and frequencies for light waves, however, differ considerably from water waves. The wavelengths of most water waves can range from millimeters to many meters. But light that is visible to the human eye has a wavelength on the order of a millionth of a meter. This is an exceedingly short length by workaday standards. About a thousand such wavelengths would fit into the width of a period in this print.

Again, for visible light the *frequency* is a few hundred million million oscillations per second, and this is vastly greater than for water waves.

The *velocities* of light waves and water waves also differ greatly. In a shallow tank of water, surface waves move with a velocity of roughly one meter per second, as is verifiable by observation. But light waves in vacuum travel at the enormous speed of 300 million meters per second or about 186,000 miles per second.

THE AMPLITUDE OF LIGHT WAVES

The amplitude of a water wave was discussed in the last chapter. Since contrasting quantities that pertain to light waves and to water waves have been discussed so far, you may well expect now a contrast between the amplitudes of light and water waves to be made.

Indeed there is a contrast, but it has to do with a more fundamental difference than just that of magnitude. The amplitude of a water wave is the vertical distance from the normal level of the surface to the crest (or down to the valley) of a wave. The amplitude of a light wave is of a different physical nature. Light does not consist of a spatial displacement in a material medium. Rather, it consists of traveling electromagnetic fields which oscillate as they move.

Understanding the electromagnetic nature of light requires substantial background in the physics of electric and magnetic fields and related phenomena, and considerable mathematics as well. Fortunately, the Maxwell electromagnetic theory of light will not be used in this book. Rather, a wave theory of light is used that is more readily understandable and more easily applied.

In the early 1800s, long before the electromagnetic theory was devised, a wave theory was originated by Thomas Young, an English scientist. This theory does not depend on knowing what light really is, but only on the assumption that it consists of transverse waves, as do water waves, and that the basic wave phenomena occur in much the same way in both cases. This simple theory of the behavior of light is completely satisfactory for understanding all of the aspects of holography that are discussed in this book.

AMPLITUDE, BRIGHTNESS, AND INTENSITY

Figure 3.1 is a reproduction of Figure 2.4, where it referred to water waves. Part (b) shows a side-on view of a wavetrain. This is not to be considered a literal representation of a light wavetrain. Nevertheless, to think of light waves in this

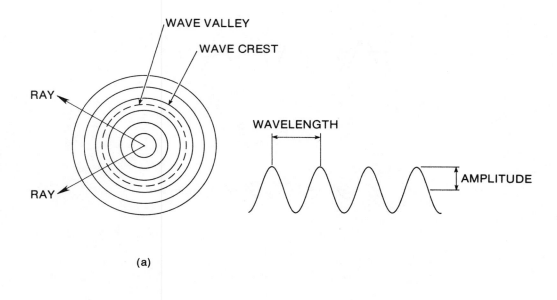

(a)

(b)

Figure 3.1. *(a) Wavefronts from a point source of light. (b) The meanings of "wavelength" and "amplitude."*

pictorial fashion is very useful because it helps the analogy of light with water waves.

Everyone knows what is meant by the "brightness" of light, but it is actually a complicated matter of physiology involving the eye and much of the central nervous system. How bright a sample of light is varies from person to person and according to the circumstances, such as competition with ambient illumination. Brightness is not a basic and inherent property of light.

On the other hand, the "intensity" of light *is* a physical property of the light itself. It is related to the amplitude of the light wave. Intense light is light with large amplitude, and less intense light has lesser amplitude. Intensity is independent of any observer. While high intensity means great brightness ordinarily, the "brightness" is a subjective matter. You might expect that the intensity is proportional to the amplitude but, in fact, the intensity is proportional to the square of the amplitude. Thus, if one wave has twice the amplitude of another, the first has four times the intensity of the second.

To illustrate these ideas, parts (a) and (b) of Figure 3.2 are presented. The waves in (a) and (b) have the same wavelength but the light in (a) is much more intense than that in (b). In fact, since the ratio of the amplitudes is about 2, the ratio of the intensities is about 4 (the square of 2).

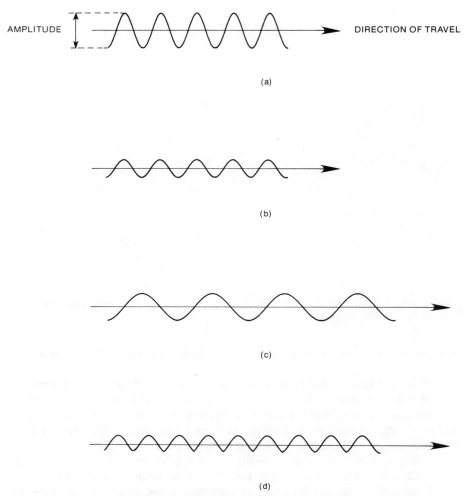

AMPLITUDE ↕   DIRECTION OF TRAVEL

(a)

(b)

(c)

(d)

Figure 3.2. Representation of some properties of light waves. (a) A wavetrain. (b) Another, with same wavelength but lesser amplitude, and hence less intensity. (c) Same intensity as in (a) but longer wavelength. (d) Same intensity as in (b) but shorter wavelength.

COLOR AND WAVELENGTH

The relationship between color and wavelength is somewhat like that between brightness and amplitude. As with your perception of the brightness of light, your sensation of color is the result of a complicated physiological process. It may vary from person to person and according to circumstances. Colorblind people do not perceive colors as other people do, for example. However, the wavelength of a light wave _is_ an intrinsic physical property of the wave itself.

The normal eye is so constituted that it responds to light only in a limited range of wavelengths, namely from about 0.4 to 0.7 millionths of a meter.[1] Light at the low end of that range is perceived as violet and light at the high end is perceived as red. If the wavelength were to vary continuously from the low end to the high end of the range, you would see the color change from deep violet through shades of blue, green, yellow, orange, and red.

A particular instance that will be important in this book is the case of light from the helium-neon gas laser. This light has wavelength 0.6328 millionths of a meter or 0.6328 microns, or 6328 Angstroms, and appears to the eye as red.

Each of the light waves represented in Figure 3.2 is especially simple. Each has a geometric shape called "sinusoidal," each has the same amplitude along its extent (and so would appear to have unvarying brightness), and each has the same wavelength everywhere (and so would appear to have an unchanging color). Because it has a sharply defined and unchanging wavelength, each such wave would be said to be _monochromatic,_ which means "of one color." Such light is exceptional in daily living, for the light to which you are ordinarily exposed is a complex mixture of wavelengths and, hence, of colors. However, monochromatic light is of basic importance in holography, as you will see.

Let us look at Figure 3.2 again to review some of the ideas that have been discussed. The light in (a) is more intense than is the light in (b), but the two have the same wavelength. If the wavelength were 0.58 millionths of a meter, the color perceived would be yellow-orange, much like the color produced by sodium vapor highway lights.

The light in (c) is as intense as that in (a), but has a longer wavelength. If this were 0.63 millionths of a meter, it would appear as red. The light in (d) is approximately as intense as that in (b), but it has a wavelength about four-fifths as great and would appear as blue.

THE INTERFERENCE OF LIGHT WAVES AND COHERENCE

Light waves superimposed on each other interfere much as water waves do. In particular, we can now interpret Figure 2.6 as a representation of sets of circular wavefronts of light spreading out from the point sources shown. Figures 2.7 and 2.8 also can be reinterpreted in terms of light waves.

---

[1]There are numerous units in use in specifying wavelengths. We adhere nearly always to the meter. One millionth of a meter is also known as a "micron." Especially common is use of the "Angstrom," which is 10^{-10} meters.

Details of the interference pattern are discussed in the next chapter, but first let's discuss more fully the ideas of coherence and incoherence.

Originally, when Figure 2.7 represented water waves, the ripple tank apparatus drove two dabbling points in synchronization and with perfect regularity. Suppose instead that the two points were driven by *separate* motors. If the two motors were exactly alike, perfectly synchronized, and never faltered, the interference pattern would be produced as before. But if one of the points, let's say B, behaved in an erratic fashion, touching the water too soon or too late, the interference at any point on the surface would not always be of the same nature. For example, at point *a* the A waves would pass by with perfect regularity but the B waves would not. When a crest in the A waves occurs at point *a,* a crest in the B waves may not be there. There might even be a valley in the B waves at the point. The fluctuations between constructive and destructive interference eventually would spoil the interference pattern. If *both* sources were erratic, the situation would be even more chaotic. The sources (and hence the waves) would then be *incoherent.*

A steady interference pattern can result only if the superimposed waves have a relationship that is the same at all times at all points. The construction of the ripple tank apparatus enforces this condition. It is not possible to do this with ordinary light sources such as incandescent lamps. To see why this is so, refer again to Huygens' drawing in Figure 2.1. Points in the candle flame are not synchronized sources and each emits in fits and starts.

Similarly, the filament in an incandescent light bulb emits incoherent light because the light is produced by individual atoms. Emissions occur erratically for any one atom and without synchronization between different atoms. When light from a lamp falls on the page of a book, you do not see the alternating dark and light regions that would be seen if steady interference took place. The pattern is "washed out" by the randomness in the light waves. Only a general averaged level of illumination of the page is seen.

Making a hologram requires that two beams (the reference beam and the object beam) fall on the plate and that they produce a stationary interference pattern. If the pattern is not to be washed out, *coherent light must be used.*

In practice, the only good source of coherent light is the laser. In a laser the atoms in the medium (solid as in a ruby laser or gaseous as in a helium-neon laser) are forced to emit in step with each other, and also all the atoms undergo the same energy transitions at each emission and so emit light of the same color. The result is that the light in the laser beam is highly coherent.

The laser is the coherent source *par excellence,* but it is not the only source with some coherence. In fact, even ordinary light sources such as incandescent lamps have a little partial coherence, though not nearly enough for making holograms.

Consider the light waves in Figure 3.2 and, in doing so, imagine that the limited segments shown actually continue indefinitely far to the right and to the left. Each of the wavetrains is perfectly uniform with respect to wavelength and frequency all along its length. In contrast, look at the wavetrains shown in Figure 3.3. In (a) the wavelength is different in the leading and in the following parts of the wave. In (b) the wavelengths in the two parts are the same, but the crests in the leading part are not in step with the crests in the following part. In (c) the light occurs in bursts, with intermittent periods of no light.

The waves in Figure 3.2 all have *temporal coherence*. A wavetrain has temporal coherence if the crests are in step with each other all along the wave. In contrast, those in Figure 3.3 lack temporal coherence, since for one reason or another each fails to meet the in-step condition.

Figure 3.3. Ways in which light waves can fail to be temporally coherent. (a) Change of wavelength. (b) One wavelength, but with a phase shift along the wavetrain. (c) Intermittent interruptions of the light.

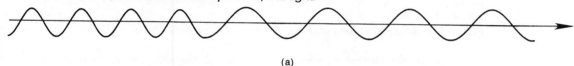

(a)

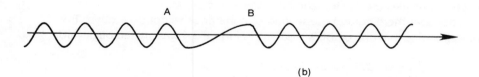

(b)

(c)

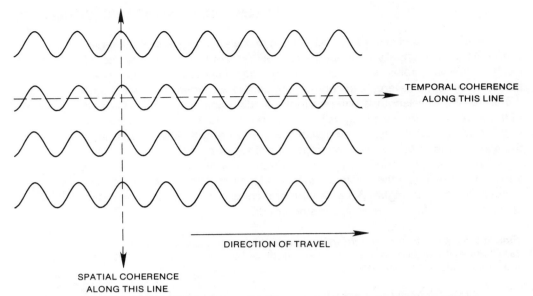

TEMPORAL COHERENCE
ALONG THIS LINE

DIRECTION OF TRAVEL

SPATIAL COHERENCE
ALONG THIS LINE

Figure 3.4. *Illustrating temporal and spatial coherence in a beam of light.*

Temporal coherence or incoherence has to do with conditions *along* a wave; *spatial coherence* refers to conditions *across* a light beam. The beam has spatial coherence if all parts are in step across the beam. Along such a ray as that indicated in Figure 3.4 by the single-headed dashed arrow, there is temporal coherence. Along the double-headed vertical arrow, the in-step condition is met and there is spatial coherence. In terms of the soldiers mentioned in Chapter 1, all files are in step and all ranks are in step.

Notice that there could be one kind of coherence without the other. For example, imagine that one of the wavetrains in Figure 3.4 was shifted backward

Figure 3.5. *A case in which a point source of light produces* spatially *coherent light without* temporal *coherence.*

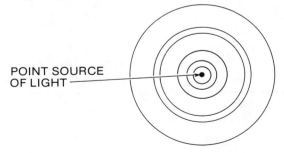

POINT SOURCE
OF LIGHT

enough to spoil the spatial coherence. In spite of this, there would still be temporal coherence in the beam.

It is also possible to have spatial coherence without temporal coherence. In Figure 3.5 the point source is creating successive Huygens' wavelets somewhat sporadically, so that the circular wavefronts are not spaced uniformly. Along any one ray there is not temporal coherence. Nevertheless, along any circle centered on the source all parts of the wave are in step, so there is spatial coherence.

POINT SOURCES, MONOCHROMATIC SOURCES, AND COHERENCE

By considering the case just discussed, you can see that a point source of light would be a source of spatially coherent light. Unfortunately, there are no real true point sources of light. However, the principle is a useful one to bear in mind. An example of its use is in converting an ordinary source such as an incandescent bulb into a nearly spatially coherent source. To do this, you can use an opaque screen with a tiny hole in it. The smaller the hole, the better the coherence, but also the smaller the hole, the more drastic the reduction in the amount of light available for use.

Some kinds of holograms (reflection holograms in particular) can be viewed with an ordinary light source rather than with a laser. In such cases, small—that is, approximately pointlike—sources are best. Also, clear and not frosted bulbs are best because the small irregularities over the surface of a frosted bulb tend to make it act effectively like a source larger than the incandescent filament.

Monochromaticity (purity of color) and temporal coherence go hand-in-hand. You can say that the monochromaticity of laser light is what gives the light good temporal coherence, or you can claim the converse relationship. Also you see readily why an ordinary white light source is not a temporally coherent source.

This principle explains a simple but effective artifice much used in holography, but mostly in *viewing* rather than in making holograms. It consists of putting a color filter in front of a white light source to make the beam more pure in color. This again reduces the intensity of the light available for use, and in practice a slide projector is often used because of the high initial intensity it provides.

In most vicinities you can find sodium lamps. These are often used for lighting highways because they emit mostly in a narrow yellow region, and this kind of light penetrates fog and mist well. Those provide interesting approximately monochromatic sources for viewing holograms. In an educational institution, the physics department probably has a mercury vapor lamp. Such a lamp emits light that can be filtered to give reasonably intense and monochromatic blue light. In fact, the world's first holograms were produced by Dr. Gabor with such a source long before lasers existed.

Holography:
The geometric model

The physics of holography is well understood today. The complete theory is, however, highly mathematical. It begins by representing light waves with certain abstract expressions and then explores mathematically the interference and other phenomena that arise.

Fortunately, it is possible to extract from the elaborate mathematics of the theory intuitively understandable views of holography. The purpose of such models is to offer simple ways to think pictorially of phenomena, though with sacrifice of completeness.

In fact, *two* good models of holography have been devised—good because they are easy to grasp and because they give insight into how holograms work. They are so useful even specialists in holography rely heavily on them.

One of these is the *zone plate* model. This model has been used in popular and technical discussions of holography since the time of Dr. Gabor's original invention of the subject. It is a powerful model and easily understood once you have learned about a class of optical effects known as "diffraction" phenomena.

An even simpler model of holography was devised by Dr. Tung H. Jeong of Lake Forest College in Illinois.[1] For many years Dr. Jeong has been a leader in holography. His *geometric model* is especially easy to understand, though limited in certain ways in its applicability. Because it is so clear and so effective where it does apply, the geometric model will be discussed first, leaving the zone plate model for a later chapter.

PHOTOGRAPHING THE INTERFERENCE PATTERNS

Chapter 3 discussed how light waves from two coherent sources interfere. The interference produces a stationary pattern of constructive and destructive *fringes*. This is shown again in Figure 4.1. The role of the photographic plate shown will be examined later.

The fringes as shown in the sketch are cross-sections of what are actually three-dimensional surfaces.[2] You can visualize the surfaces by thinking of rotating Figure 4.1 about a line through A and B. For clarity, part of one of the surfaces is shown in Figure 4.2, again intercepted by a plate. At all points on this surface the intensity of the light is at its greatest due to constructive interference. The next neighboring surface (not shown) also consists of points where there is constructive interference, and between the surfaces there is low intensity of light, due to destructive interference.

Figure 4.1. For making a transmission hologram a photographic plate has been placed in the interference pattern that is produced by the coherent sources A and B.

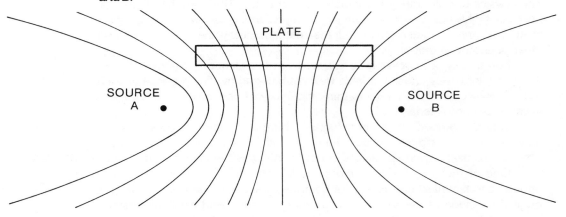

---

[1]Tung H. Jeong, "Geometric Model for Holography," *American Journal of Physics*, August 1975, pages 714–717.
[2]As a matter of terminology, the three-dimensional surfaces are *hyperboloids* or, more properly, *hyperboloidal surfaces*, although they are often simply called hyperbolas.

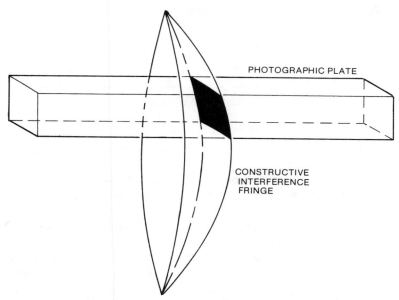

Figure 4.2. *Part of one of the hyperboloidal surfaces in the interference pattern. The segment shown as dark activates the corresponding part of the emulsion.*

An essential step in holography is producing the stationary fringe pattern. The next step is recording the pattern to form the hologram. The emulsion in Figures 4.1 and 4.2 reacts strongly where the light is intense and less strongly elsewhere. Thus the constructive interference fringes will become surfaces where there has been strong activation of the silver halide grains in the emulsion. After the exposed plate has been processed chemically, it will be a photographic negative that contains a permanent record of the interference pattern.

The next step in holography is to use the plate to reconstruct a visible decoded version of the information that was recorded. Before discussing this step, let's consider certain properties of the hyperboloidal surfaces further.

THE FRINGES AS REFLECTING SURFACES

Figure 4.3 shows the nature of the photographic negative in a cross-sectional view. Each of the fringes is now to be thought of as a *partially reflecting surface.* Suppose you were to shine light on the plate. Where the incident light falls on one of the surfaces, some of it will be reflected back, some will be transmitted through, and some will be lost by absorption. The general nature of the reflection from one of the surfaces is shown in Figure 4.4. Light from the real source A

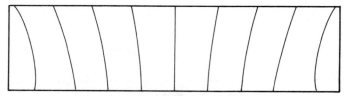

Figure 4.3. *The nature of the interference fringes that are recorded in the plate.*

reflects in the manner shown to the observer's eye. (The "virtual source B" will be discussed presently.)

The whole plate is to be thought of as a set of nested partially reflecting surfaces, each of them acting like the one in Figure 4.4 when the plate is illuminated.

The importance of the hyperbolic surfaces lies in the precise way in which they reflect light rays. Consider Figure 4.5 (a). The curve shown represents one of the partially reflecting surfaces, originally made using real sources A and B. Now the source B has been removed, and only the source A illuminates the plate. The ray R_1 reflects from the surface as ray R_2. Similarly ray R_3 reflects as

Figure 4.4. *The manner in which rays from a real source of light (A) reflect from one of the partially reflecting surfaces in the plate.*

OBSERVER'S EYE

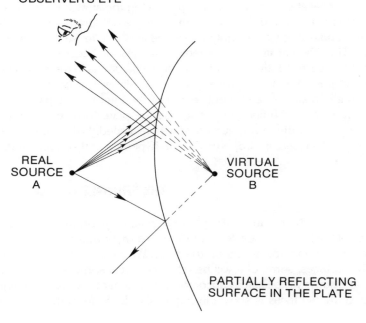

REAL
SOURCE
A

VIRTUAL
SOURCE
B

PARTIALLY REFLECTING
SURFACE IN THE PLATE

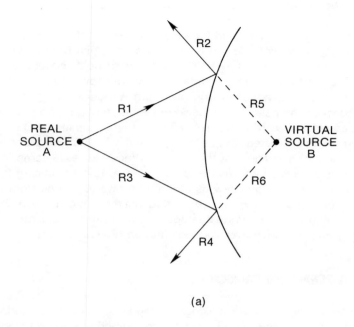

(a)

Figure 4.5. *Rays from the point source A seem after reflection to have come from a virtual point source B. (a) Convex surface. (b) Concave surface.*

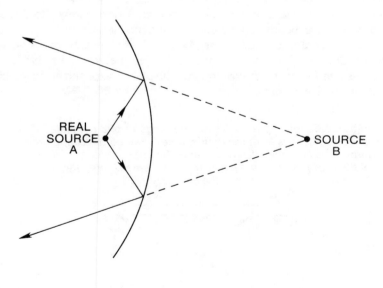

(b)

ray R_4. If the reflected rays R_2 and R_4 are traced backward, as with the dashed lines R_5 and R_6, these extensions will intersect at the point B. In part (a) of the figure, the surface is convex toward A. The theorem holds equally well in the case shown in part (b), where the surface is concave toward A.

Furthermore, this holds for *any* rays from A. Hence in Figure 4.4, all of the rays shown as emanating from the source A and reflecting from the surface will, when traced backward from their reflected parts, intersect at B.

If you see the reflected rays as in Figure 4.4, what will you perceive? The eye has no way of knowing that the light rays received really came from the source A after reflections and that there are no real rays emanating from the point B. To the eye, it is precisely as though the rays *did* come from B. THis is called a *virtual image,* thereby emphasizing that the source is not really there. As pointed out in Chapter 1, virtual images are far from being esoteric optical entities; they are seen whenever you look into an ordinary mirror.

A HOLOGRAM OF A POINTLIKE OBJECT

At last we are ready to discuss making a hologram of something and viewing it. We do this in two stages. In this section the "something" to be holographed will be only a very small, almost pointlike object. In the next section, the ideas that will have been developed will be applied to the general case—a hologram of an extended scene.

In Figure 4.6 some of the light from source A goes directly to the plate; this is the reference beam. Some of the light (not indicated by rays in the drawing) goes from the source A to the object P. This reflects from P, and some of it falls on the plate. The latter light is the object beam. The reference and object beams interfere in the emulsion. After the plate is exposed and processed, it is a hologram of the pointlike object.

Figure 4.6. *Making a transmission hologram of a pointlike object P. The rays from the light source A that go directly to the plate constitute the reference beam. The object beam consists of the rays from P to the plate. These are created by reflection by the object P of light from the source A.*

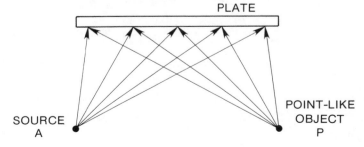

PLATE

SOURCE
A

POINT-LIKE
OBJECT
P

It would be well to compare Figures 4.1 and 4.6 at this point. Although in 4.1 the source B is self-luminous, while in 4.6 the object P shines only by reflected light, yet the geometry is the same in the two cases. That means that the *shapes* of the fringes and their *locations* in the emulsions are the same in the two cases. Hence with reconstruction as in Figure 4.7 a pointlike virtual image will be seen in either case, and in the same apparent location. However, how the virtual image looks depends on still another property of the fringes— their degrees of blackness or their contrast with the generally more transparent background in the emulsion. If the reflecting object P is so shiny that it reflects almost all of the light that falls on it, the interference of the reference and object beams in the emulsion will produce fringes with high contrast. When the image is reconstructed, it will be bright. On the other hand, if the original object P is dull, reflecting little light, the fringes in the emulsion will be of low contrast, and the image will be dull.

This is of fundamental importance, for it is the basis for understanding how the holographic image of a scene can show varying degrees of brightness in various parts of the scene, as is required for a realistic representation of the original scene.

Besides this matter of "dynamic range," the realism that is perceived in the reconstructed image depends on having all the points in the scene in their proper relative positions in three-dimensional space.

Figure 4.7. *Viewing the hologram of a pointlike object. The reconstruction with light from the source A gives the viewer the impression of seeing the object P, though this is not actually present.*

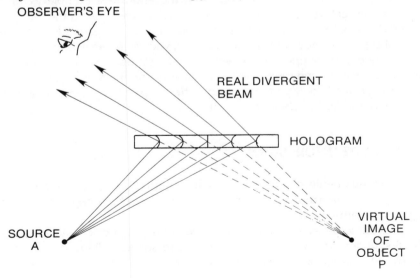

OBSERVER'S EYE

REAL DIVERGENT BEAM

HOLOGRAM

SOURCE A

VIRTUAL IMAGE OF OBJECT P

HOLOGRAM OF AN EXTENDED SCENE

If, in making the hologram as in Figure 4.6, the object P had been located at a different point (nearer to or farther from the plate or displaced in some other fashion), the fringes recorded in the hologram would be different from before and in a manner corresponding to the displacement. On reconstruction of the virtual image as in Figure 4.7, the viewer would perceive the object as though it were located where it was originally. Thus the plate contains information about position in the three-dimensional space.

Figure 4.8 shows how a transmission hologram of a sizable object can be made. The coherent light needed is a beam from a laser. The beam is diverged by a lens to spread it out enough to illuminate the object. In the figure, four points on the surface are shown as reflecting light. Sets of hyperbolic surfaces will be formed by interference, one for each part of the surface. Each set will differ in detail from the others.

When the hologram is viewed, the four parts of the surface will be seen, each in its proper spatial position with respect to the others and each with its proper brightness or dimness. In short, they will be seen recreated realistically.

At this point a mental leap is required. You are to think of *all* the points of the object as behaving in the same way as our four representative points. The hologram of the object will contain a very complicated collection of sets of interference fringes. However, when the hologram is illuminated for viewing, the inexorable laws of reflection of light rays take over. The various sets of fringes will produce images of the different points, all properly sorted and placed in their correct spatial relationships. The scene will have true depth.

The arrangement for reconstructing the scene is illustrated in Figure 4.9. (No attempt has been made to cope with the enormous complexity of the fringes in the real situation.) Notice that if you put your eye to the left in the "viewer's location" region, you will see the object from a corresponding direction and similarly if you move to the right. In short, you will be able to look part way *around* the object. Suppose there were two objects used in making the hologram, one partly behind the other. On reconstructing the scene, you could, by moving your head, look part way around the image of the nearer object and see parts of the farther object come into view.

PHASE AND AMPLITUDE INFORMATION

The most fundamental characteristic of a hologram is that it records both *phase* and *amplitude* information. In contrast, an ordinary photograph records only amplitude information.

In order to understand this much used but highly succinct statement, you must first understand what is meant by phase. Imagine two light waves traveling side by side, having the same wavelengths and the same amplitudes as any two of those shown in Figure 3.4. All of those shown in the drawing are also in step across the beam. Now concentrate on the uppermost two of the waves and think of the second of them as shifted to the right. After this shift, the two waves are still identical with respect to wavelength and amplitude (or color and intensity), but they differ in *phase*. To become more precise about this, Figure 4.10 shows five waves, all the same except for phase shifts relative to each other. Take A to be one with respect to which the shifts of the others can be measured. One way to specify the amount of shift of any of the others is to state it as a fraction of one wavelength of the light. This will be the *phase* of the wave relative to the standard A. The phase of B is 0 and the phases of C, D, and E are approximately 1/8, 1/4, and 1/2 wavelength, respectively.[3]

In making a hologram, a reference beam and an object beam interfere. For constructive interference, the beams are said to have zero relative phase, or that they are *in phase*. For destructive interference, the beams have relative phase 1/2 wavelength. They are completely *out of phase*. For intermediate phase differences, the interference is partial. What is recorded in the photographic emulsion is the interference pattern, and this depends at every point upon the *phase relationship* between the reference and object beams. Therefore, when the emulsion records an interference pattern, it is truly recording phase information.

What about amplitude information *from the scene?* This question can be distinguished by the way in which phase information is converted into amplitude information when the interference occurs in the plate, and by amplitude information about the scene carried by the object beam itself, before the object beam encounters the reference beam in the plate. After all, if you look at something in ordinary light, you detect differences in brightness in various parts of the scene, and this is not a matter of interference with a reference beam. Amplitude information about the scene in this sense must also be recorded in a hologram. In short, how does the hologram know what parts of the scene are bright and which are dim?

In making a hologram as in Figure 4.8, the various parts of the object being holographed have various reflectivities, various intensities of light falling on them, and are at various distances from the plate. All of these factors contribute to determining how intense the object beam is when it meets the reference beam at each point in the plate. A weak object beam somewhere means weak interference, and a strong object beam elsewhere means strong interference there. Thus the degree of blackening of an interference fringe in the plate depends on the amplitudes of light from various parts of the object as well as on phase. A hologram records this information and so can be said to contain amplitude information about the scene.

In making an ordinary photograph (with ordinary light, of course), there

---

[3]Another way is to consider one wavelength to be equivalent to 360 degrees. With this convention, the phases of B, C, D, and E relative to A are approximately 0, 45, 90 and 180 degrees.

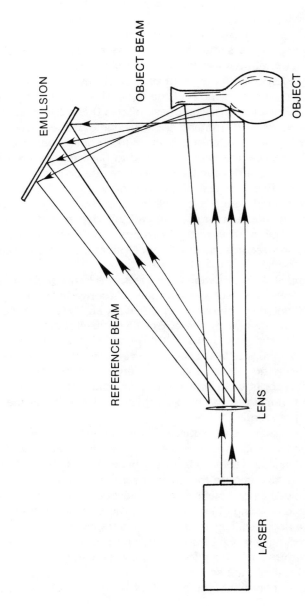

Figure 4.8. *Making a transmission hologram of an extended object.*

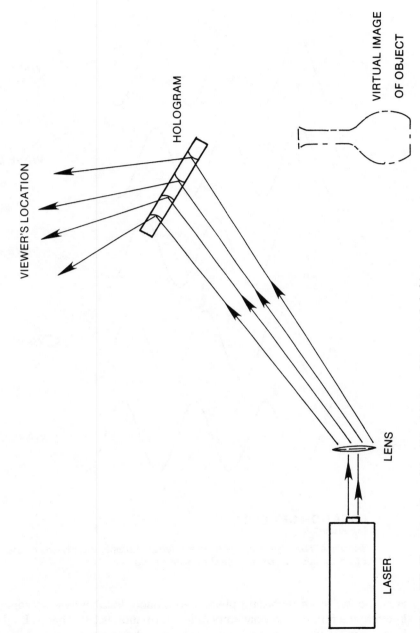

Figure 4.9. Reconstruction of the virtual image of the extended object.

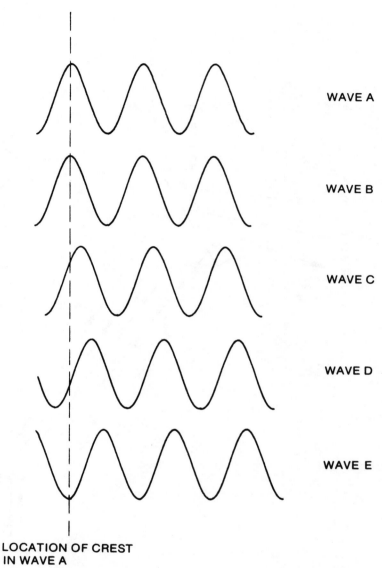

WAVE A

WAVE B

WAVE C

WAVE D

WAVE E

LOCATION OF CREST
IN WAVE A

Figure 4.10. *Using wave A as a reference wave, the phases of the other waves can be specified quantitatively.*

is no possibility of recording phase information. Incoherence smears out the interference fringes and only amplitude information is left. That is, bright spots in the scene activate the emulsion in corresponding parts more than do dimmer spots in the scene, and this is purely amplitude information.

44

Figure 4.11. *The general nature of a portion of a hologram, as it would appear under a microscope.*

THE REALITY OF THE PARTIALLY REFLECTING SURFACES

In using the geometric model, you may think of the hyperboloids as partially reflecting surfaces. At the same time, it is going too far to take the model so literally as to suppose that the surfaces are continuous, shiny surfaces like our familiar silvered mirrors. For one thing, photographic emulsions are necessarily *grainy,* and this graininess limits their ability to contain continuous surfaces in them. Also, when a scene is holographed, the exceedingly numerous points in the scene produce innumerable overlapping fringe systems, with a resultant complexity so great that even in the absence of the graininess effect a microscopic view of the plate could hardly be expected to reveal distinct continuous surfaces. In fact, a hologram seen under a microscope appears as a collection of squiggles, somewhat as in Figure 4.11.

That this is so may seem to make it miraculous that the geometric model works at all. But if the pieces of mirrors (the squiggles in the hologram) have sufficient organization among them, as they do, various sets of pieces can act consistently as fragments of continuous mirrors. As an analogy, you might think of a mosaic of pieces of silvered mirrors embedded in plaster and so cunningly put into place that from various viewpoints you can see your face reflected.

REFLECTION HOLOGRAMS

A transmission hologram is viewed with the light source and the eye on opposite sides of the plate, so that the light is transmitted through the plate. A *reflection hologram* is viewed with the light source on the same side of the plate as the observer's eye. The light reflects from the plate to the eye, thus the name is appropriate. Since every hologram is illuminated from the one side or the other, every hologram is of one kind or the other. (However, there are numerous varieties of each kind, as you will see later.)

As a first step toward understanding reflection holograms, consider the arrangement shown in Figure 4.12. A and B are again point sources of coherent light and they produce the hyperbolic interference fringes. The new feature in the arrangement is that the photographic plate is oriented differently from how it would be if a transmission hologram were being made. The plate now has the interference fringes extending essentially lengthwise along the long dimension of the plate. The nature of the hologram formed is represented schematically in part (b) of the figure.

As a second step, suppose that point A is a source of coherent light, but that B is a small (pointlike) reflective object. The plate will receive a reference beam from A and an object beam of reflected light from B. A hologram of the object will be formed in the plate, for the emulsion will react to the varying

Figure 4.12. *The plate is placed in a position for making a reflection hologram. (a) The interference fringes lie essentially along the length of the plate. (b) The general nature of the fringes ultimately recorded in the plate.*

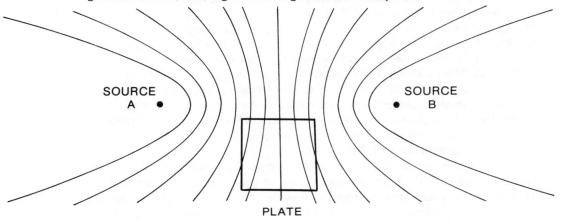

(a)

(b)

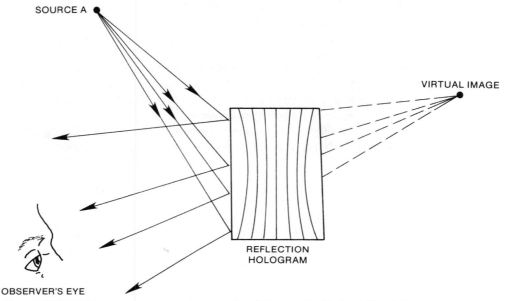

SOURCE A

VIRTUAL IMAGE

REFLECTION
HOLOGRAM

OBSERVER'S EYE

Figure 4.13. *Reconstructing the image with a reflection hologram. The light source and the viewer's eye are on the same side of the plate. For simplicity the rays are shown as if they reflected from the face of the plate, but in reality they reflect from the surfaces that are shown in the plate.*

intensities of light in the interference pattern. The geometry of the fringes recorded in the hologram will be different from a transmission hologram, however.

As a third step, observe that the geometric model provides for reconstruction of the virtual image of the small object. This is illustrated in Figure 4.13. Light from the source A reflects from the surfaces in the hologram and back to the eye. You interpret what you see as though the light rays came from an object at B.

As a last step, imagine the point B as but one point of the very many points that make up the surface of an extended object. Every point will give rise to fringes in the hologram so that when the hologram is viewed each point will be perceived in its proper three-dimensional relation to all of the others and with appropriate brightness. The viewer will interpret the ensemble of the impressions received as a realistic virtual image of the object.

THE STRATEGY FOR THE NEXT THREE CHAPTERS

In this chapter the geometric model has been used to explain much about transmission holograms but our story is incomplete. For one thing, only virtual images, not real images have been discussed. The latter topic is somewhat

47

elaborate and will be taken up in Chapter 5. A number of other properties of transmission holograms will also be examined.

This chapter also introduced reflection holograms, but there is much more to be said. In particular, our introduction did not deal with an optical effect called Bragg reflection, which is predominant in reflection holograms. This is very important, especially in connection with color effects. Since it requires some preliminary delving into optical theory, Chapter 6 takes this up, along with the general topic of reflection holograms.

Chapter 7 uses the enlarged theory (geometric model plus Bragg reflection) to examine both transmission and reflection holograms further.

More about transmission holograms

Chapter 4 explained in an introductory way how the geometric model is used to understand the formation of images in holograms. This chapter applies the model to transmission holograms in particular. You will see that many of their specific properties can be explained simply now.

There is a highly important exception—color effects. Wouldn't it be wonderful if transmission holograms could be viewed with ordinary white light sources such as sunlight or electric lamps, doing away with the need for a laser? Wouldn't it also be wonderful if the images were produced in full natural color? Unfortunately, accomplishing these things is not yet in the realm of practical reality.[1] For example, if a transmission hologram is viewed in ordinary white light, the image is seen as a smear with a spectral distribution of colors across it, rather than sharply in natural colors.

This is shown in Figure 5.1. Bear in mind that until the last section of this chapter it will be assumed that the images are reconstructed with laser light of the same kind as was used in making the holograms, or with nearly monochromatic light of the same color derived by using a color filter with a white

---

[1]We speak here of ordinary transmission holograms. Special types such as "rainbow" transmission holograms, which are white-light viewable, are discussed in later chapters of this book.

Figure 5.1. *Here the same hologram as in Figure 1.1 is illuminated with a white light source. The smears seen in this black-and-white photo would appear as spectral color smears were the image viewed "live." (Photo by Dr. C. E. Hamilton, produced by Mr. Rainbows.)*

light source. Under those conditions, the geometric model gives the right answers and is, in fact, very impressive in its explanatory power. The last section of this chapter will comment on the nature of the color problem and indicate how it will be dealt with in this book.

A GENERAL DESCRIPTION
OF THE REAL IMAGE

In Figure 5.2 a diverged laser beam lights the hologram from the left. The hologram "processes" the light and *three* beams emerge. One beam diverges and travels upward. This beam produces the virtual image. Another beam converges, or focuses, and travels downward. This beam produces the real image. There is also an undeviated beam. This "direct" beam gets through the plate as a kind of leakage. It does not carry holographic information about the scene and therefore will not be discussed further.

If you look into the divergent beam you are able to see the virtual image because the lenses in your eyes provide the requisite focusing action. On the other hand, the convergent beam produces the real image without the need for any external lenses. A simple and interesting experiment consists of letting the beams fall on a white card or piece of paper. If the card is placed in the divergent beam, only a vague region of unfocused light will be seen. If the card

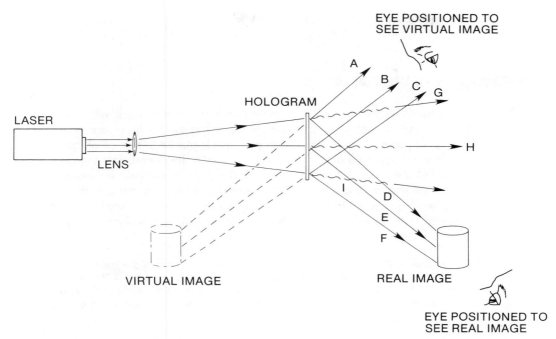

Figure 5.2. *A transmission hologram producing two images. Rays A, B, C pertain to the virtual image. Rays D, E, F pertain to the real image. Rays G, H, I pertain to the direct beam.*

is placed in the convergent beam, the objects in the scene *will* be imaged. This is the real image.

Since we are assuming that the readout light has the same color as that used to make the hologram, both images will be seen in that color. Ordinarily this means the images are in the characteristic red color of the helium-neon laser.

The real image has a number of special properties. One of these is that it exists over a considerable region of space. If you move the card closer to or farther from the hologram (perhaps even over a range of several feet), the image will still appear on the card. This means that the *depth of focus* is very large.

With the card in any position, the objects in the scene can be distinguished, but some will be more sharply in focus than others. As the card is moved, the in-focus parts change. Different parts are in focus at different distances from the plate but the card can show you only two-dimensional slices.

With the method of reconstructing the image shown in Figure 5.2, the objects in the image may be seen greatly magnified in comparison with the sizes of the original objects. For example, an object that was one inch tall may appear as one foot tall.

Finally, if you place your eye in position to see the real image with or without the card as an intermediary, you see a very curious property of the image.[2] The effect is called *pseudoscopy*.

PSEUDOSCOPY IN THE REAL IMAGE

The pseudoscopic view is best understood by contrasting it with the more familiar *orthoscopic* view of real objects. In Figure 5.3 there are two chess pieces—a pawn and a queen. When your eye is placed to the left of the dotted line as in (a), it sees the pawn as closer to it than the queen, with the queen to the left of the pawn. If the eye moves to the right, the relative positions of the pieces will appear to change, with the queen appearing to move to the right of the pawn. You should be able to see this sequence shown in the figure and act it out by holding a finger of each hand before you and shifting your point of view.

Figure 5.3. (a) The viewer's eye sees the queen to the left of the pawn. (b) The eye has moved to the right and the queen is behind the pawn. (c) The eye has moved farther to the right, and the queen is now seen to the right of the pawn. In effect, from (a) and (c) the queen "moves" to the observer's right, relative to the pawn.

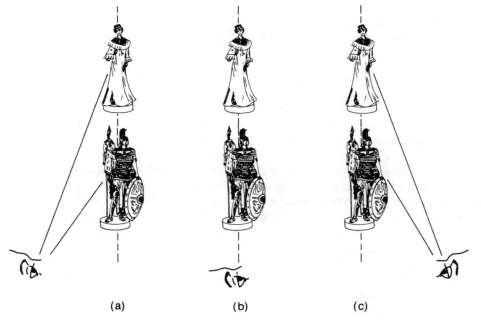

(a)	(b)	(c)

---

[2]It is best to use a card to locate the image before positioning the eye, otherwise considerably visual searching may be needed. With a low-power laser and a well-diverged beam there is no hazard in looking into the beam momentarily. The reader is urged to read Appendix I, "Laser Safety."

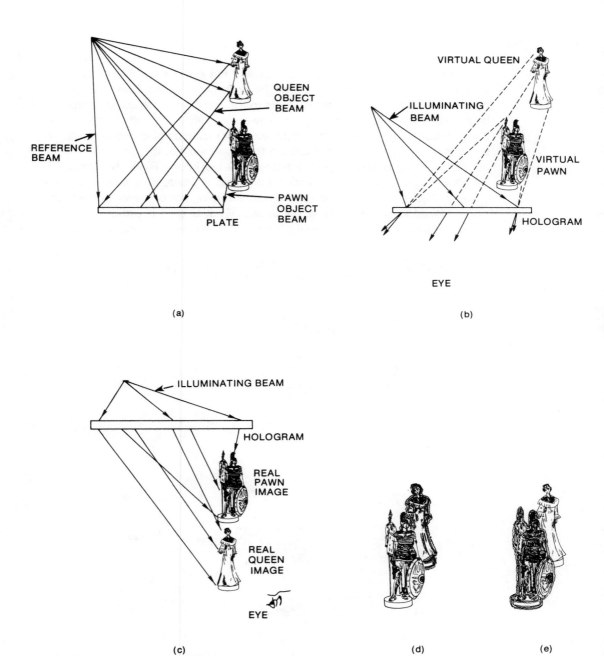

Figure 5.4. (a) Making a transmission hologram of a pawn and a queen. (b) Orthoscopic view of the virtual images. (c) Viewing the real images with the eye. (d) If the eye is near the plate, the pawn is in focus and seems nearer the viewer than does the queen. (e) If the eye is farther from the plate, the queen comes into focus, but the pawn still seems nearer to the viewer. Thus where the images really are and where they seem to be are opposite.

Notice that when the chess pieces and the eye are completely lined up as in (b), the pawn will block some of the body of the queen from view. There can be no doubt that the pawn is closer to the eye than is the queen. If the eye is shifted to the left, the queen also is seen to shift to the left relative to the pawn. These changes in the apparent left-right locations of objects are said to be *parallax* effects. This is the usual orthoscopic view of things.

Now let's consider the pseudoscopy in the real image. Figure 5.4 (page 53) (a) shows again how a transmission hologram is made, this time using the pawn and queen as objects. In (b) the finished hologram is illuminated and the eye is placed to perceive the virtual image. What is seen is orthoscopic. The objects appear located as the real objects did in (a), and with normal parallax.

Figure 5.5. *By using a card to locate the images, the pawn is in best focus nearer the plate than is the queen.*

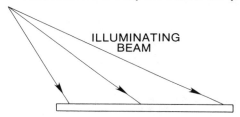

ILLUMINATING
BEAM

VIEWER'S
EYE

CARD NEAR.
PAWN FOCUSSED.
QUEEN FUZZY.

CARD FAR.
PAWN FUZZY.
QUEEN FOCUSSED.

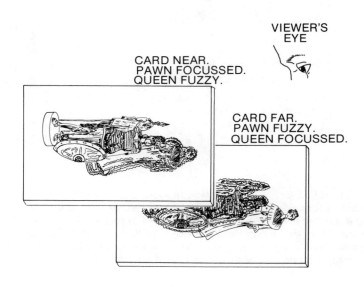

In part (c) the real image is being viewed. Remember that with the considerable depth of focus the chess pieces are not imaged as sharply localized as the drawing indicates, but rather there is a region over which the eye can be moved with the pawn and queen remaining discernible, though with fuzziness in one or the other. At one distance the pawn will be sharp, but superimposed as an obstacle on a relatively fuzzy queen. At another distance the pawn will be fuzzy and the queen sharp, yet the pawn will appear to be an obstacle between the observer and the queen. Figure 5.5 helps to clarify this. Here a card is used instead of direct viewing with the eye. As indicated, the evidence is that the pawn image is closer to the plate than is the queen image. However, if you move your eye sidewise, you will see the queen (which you know by visual evidence to be behind the pawn) move oppositely from how you would expect according to ordinary orthoscopic experience. That is, if you originally see the queen to the left of the pawn and move your eye to the right, the queen will be seen to move to the left and not to the right. There has been a reversal of the parallax effect, in this sense.

Has there also been a reversal of the near-far relationship? If the eye looks at the virtual images of the pawn and queen in Figure 5.4 (b), the pawn would be perceived as nearer the eye than is the queen. In Figure 5.4 (c) you are looking at a real image of a pawn, with the image _farther_ from your eye than is the image of the queen. Yet what you see gives the definite impression that the pawn image lies closer to you, as in parts (d) and (e). The pawn image _is_ actually farther away, and yet it seems closer, in a sense. The near-far relationship is said not to be inverted, but because of the visual evidence, you could also say the opposite. This will be clarified in the next sections.

HOW THE REAL IMAGE IS PRODUCED

We have been describing in operational terms what is meant by pseudoscopy. It is time to explain how the real image is produced according to the geometric model, and how its peculiar nature can be accounted for.

Let's begin with the various parts of Figure 5.6. Part (a) shows a now-familiar drawing. The rays from source A reflect from surfaces in the hologram and diverge. If they are traced backward, they intersect at the original location of the object at B. In part (b) the directions of the rays in (a) have been reversed. It is a basic principle in optics that the rays in (b) will behave just as did those in (a), except that the light will travel in the opposite directions. It follows that if the plate is illuminated with the converging rays, they will reflect and then converge (focus). They form the real image at A.

This is the simplest way to account for the formation of the real image. Furthermore, it is only in this way that the most sharply defined and least distorted, or "aberrated," image is reconstructed. However, the real image can be formed if the hologram is illuminated with a laser beam _diverged_ with a

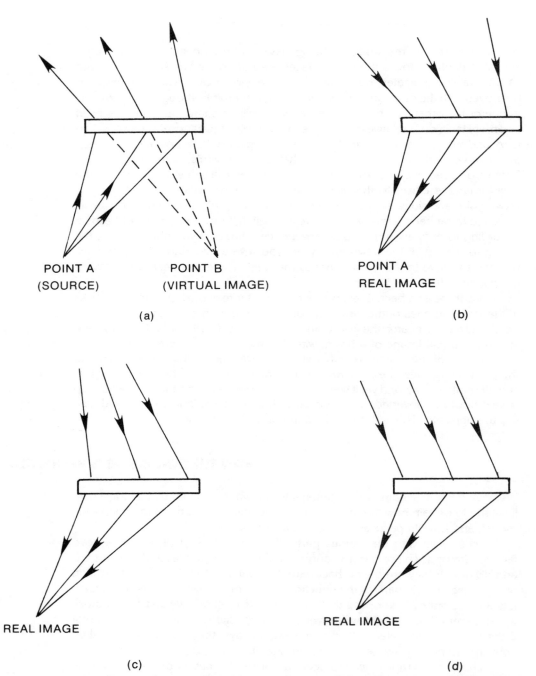

POINT A
(SOURCE)

POINT B
(VIRTUAL IMAGE)

(a)

POINT A
REAL IMAGE

(b)

REAL IMAGE

(c)

REAL IMAGE

(d)

Figure 5.6. *(a) Production of a virtual image of a pointlike object at B, using the same source A as was used in making the hologram. (b) The principle of reversal of light rays used to explain the formation of the real image. Instead of such convergent rays either divergent rays (c) or parallel rays (d) can be used.*

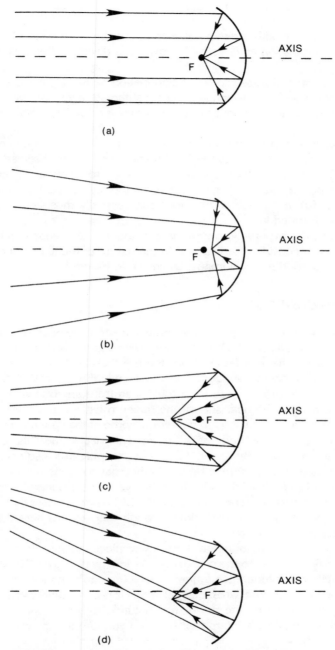

Figure 5.7. Focusing of various kinds of light beams by a concave mirror. (a) Parallel rays parallel to the axis of the mirror. (b) Convergent rays along the axis. (c) Divergent rays along the axis. (d) Convergent rays from an off-axis source.

simple lens, or even with a direct collimated laser beam. It takes great skill to arrange the illumination as described above, so a divergent beam is ordinarily used.

How can such diverse modes of illumination be used? Let's consider the following analogous case, which has the advantage that you can verify what is said if you have available a concave shaving mirror or the like.[3] See Figure 5.7 (page 57).

If parallel rays along the axis strike the mirror as in (a) they come to a focus at point F, which is called the focal point. This can be observed by using the sun or a distant street lamp and a concave mirror. This works also if the rays are not parallel, as in (b) and (c). Even if the rays come from an off-axis source as in (d), an image will be formed. Generally these images will be aberrated or distorted in some way.

Similarly, the hyperbolic mirrors in a hologram can reconstruct the real image even with considerable variation in the nature of the illuminating beam. This is also true of the virtual image, as shown in Figure 5.1.

EXPLANATION OF PSEUDOSCOPY

Pseudoscopy in the real image is visual confusion about the near-far relationship and about the parallax effect. One of its surprising aspects is that if, as in Figure 5.4 (c), the pawn-image is farther from the eye than is the queen-image, the pawn seems to be nearer. Suppose that you were looking at real chess pieces, as in Figure 5.3. In order to see all of the queen from any one vantage point, light rays from all of the near side of the queen must reach the eye. Thus in part (b) of the figure, in making the hologram, some of the light from the queen is blocked by the opaque body of the pawn. This visual clue tells us that the pawn is between us and the queen. (The sensory apparatus also uses the fact that the eye's lens muscles must be used differently to focus on the objects.)

Consider Figure 5.4 (c) again. Now the eye is not looking at real objects, but rather is reacting to light rays coming from the hologram. Rays that converge to produce the pawn-image pass right through the region where the queen-image is located, and on to the eye.

However, when the hologram of the chess pieces was made, some of the light rays from the queen to the plate *were* blocked by the pawn. The hologram cannot contain any interference fringes to which contributions were lacking in the object beam. Therefore, when the real image is reconstructed, you see some blocking of the queen by the pawn. You naturally interpret what you see as indicating that the pawn is interposed between you and the queen.

You can now observe the parallax effect if you shift your eye sidewise. You are looking at a pawn-image that is farther from you than the queen-image.

[3]Such a mirror is probably a spherical surface. The discussion refers to a mirror of a slightly different shape—a "parabolic" mirror, such as is used in reflecting telescopes.

You will see the pawn-image move with normal parallax effect but will be startled to find yourself thinking that the parallax effect has been reversed. Your surprise comes from the fact that you think of the queen as being farther away, rather than closer.

Although our example used two well-separated chess pieces, the same effects occur with a hologram of a single object. For example, a hologram of a plastic airplane model gives real images of the various points in the model, which are related as in the case of the chess pieces. If a strongly rounded object (such as the body of a flashlight) is used, you may get from the real image a sensation of looking at a concave object.

Michael Wenyon says that if you work with pseudoscopic images enough, you learn to convert what you see into orthoscopic images automatically.[4] This is reminiscent of experiments in which people wore spectacles that made the world appear upside down to them. After an initial week of confusion, the people adjusted and saw the world as right side up despite the glasses.

DEPTH OF FOCUS

The meaning of _depth of focus_[5] is well illustrated in the preceding discussions of the real image. A card placed anywhere in a range of distances from a hologram can pick up the real image. This is a case of large depth of focus. If the card had to be placed at one particular distance, the depth of focus would be small.

This subject is familiar to photographers. Even for nonphotographers the concept is easily understood if discussed in terms of a simple camera. Figure 5.8 shows representative light rays passing through a lens. In order to get a good photograph with this arrangement, the film can be placed anywhere within a certain region L without serious deterioration of the image. If the lens is wide open as in (a), the region L is narrow, but if a diaphragm is used to make the "entrance pupil" small as in (b), the region L will be wider. This "L" is the depth of focus. As indicated, it depends on the size of the lens opening. (It also depends on the ratio of the focal length to the diameter of the opening, a quantity known to photographers as the _f-number._)

When light reflects from a curved mirror the reflected rays can converge to produce a real image. The process is not the same as that by which a lens acts on light passing through it to produce a real image, of course, but the two processes are sufficiently analogous for our purposes. You can get a useful conception of why the holographic real image has large depth of focus by

[4]Michael Wenyon, _Understanding Holography_ (New York: Arco Publishing Company, Inc., 1978), pages 68–69.
[5]_Depth of focus_ is not the same as _depth of scene._ The latter has to do with how deep a scene made up of real objects can be to make a hologram of it. Depth of focus is also known as _depth of field._

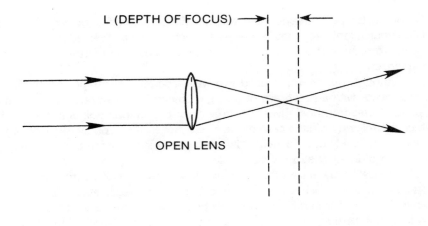

L (DEPTH OF FOCUS)

OPEN LENS

(a)

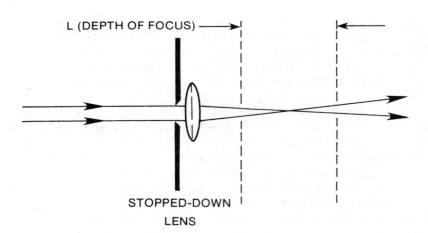

L (DEPTH OF FOCUS)

STOPPED-DOWN
LENS

(b)

Figure 5.8. *(a) A lens at full aperture has relatively small depth of focus. (b) When the same lens is stopped down it has larger depth of focus.*

thinking of the reflecting surfaces in the hologram as having large focal lengths (since they are nearly flat) and as being much "stopped-down" (since the emulsion is so thin that it contains only a small width of a surface across it). A photographer would say they have very large f-numbers.

Notice that the question of depth of focus does not arise in connection with the virtual image. In viewing this image, your eye receives a divergent beam of light and focuses to give you a sharp view, as it would if it were receiving light from an actual set of real objects. In the reconstructed scene you may pay for focusing closely on parts of the scene by losing some sharpness in the clarity with which other parts are seen, but this is true also in looking at real objects.

MAGNIFICATION

The real and virtual images of an object do not necessarily have the same size as did the original object. It is sometimes said that the virtual image perfectly reproduces objects with their proper sizes. In fact, this is not always so as can be seen by a simple experiment—if you shine diverged laser light on a hologram and then change the distance from the laser to the hologram. Changes in the dimensions of objects in the scene will be seen, and not all of these dimensions can be correct.

The _magnification_ is the ratio of the image size of an object to the actual size of the object itself. Thus magnifications greater than or less than one mean, respectively, that the image is larger or smaller than the original object. Only if the magnification is equal to one will the image show the objects in their true sizes.

There are several variables that affect the magnification: the distances of the objects and of the source of light from the plate during the making of the hologram; the distance of the source of light from the plate during reconstruction of the scene; and the wavelengths of the light used in making the hologram and of the light used in viewing.

There are two special cases in which the magnification will be equal to one. In one of the cases the light source used in making the hologram and the light source used in viewing the scene have infinite distances from the plate. That is to say, plane waves are used in both cases. This condition is not commonly met because laser light diverged by a lens is usually used.

The second special case occurs when the recording and reconstructing sources have the same finite distances from the plate. In other words, the wavefronts are spherical and alike. Obviously if you set up a hologram made with an arrangement about which you knew nothing, it is unlikely that if you illuminate it at your convenience you will duplicate the recording conditions. In the little experiment described above you can expect to see the virtual image change in size by a factor of two, three, four, or so. If the two distances are not equal, the magnification for the virtual image will always be less than one. For the real image, the magnification will be greater than one.

Up to this point, we have assumed that the wavelength of the reconstructing light and the wavelength of the viewing light are the same. This is always the case when a helium-neon gas laser is used in both instances. If different wavelengths are used, the magnification depends on the ratio of the wavelength used in viewing to the wavelength used in making the hologram, in addition to the source distances.

In principle this makes possible the achievement of large magnification. For example, if a coherent source of short wavelength X-rays were available and used in making holograms, and visible light used to view them, magnifications of the order of one million would be possible. Such sources do not exist at present so this scheme has not been brought to reality.[6] When visible light of the most extreme wavelengths possible are used, a magnification only of the order of two can be achieved in this way.

An effect that accompanies change in size of images is that they are nearer or farther from the viewer, according to the magnification. With large changes in the magnification, this can be objectionable. An example is the case of acoustic holograms, which are discussed in Chapter 10. With these the wavelengths used in making the holograms and in the viewing of them are enormously different and the final images are seen as though they were very remote.

REDUNDANCY: EACH PIECE OF A HOLOGRAM IS A HOLOGRAM

Consider Figure 5.9. The wide and narrow holograms in (a) and (b) both capture and record interference fringes from the object and each contains a holographic record of the scene. However, in (a) the plate "sees" the object from a wider

Figure 5.9. (a) A wide hologram "sees" the object from a wide range of angles. (b) With a narrow hologram, the range of angles is smaller. (c) If only a small part of a wide hologram is used, that part acts like the narrow hologram in (b).

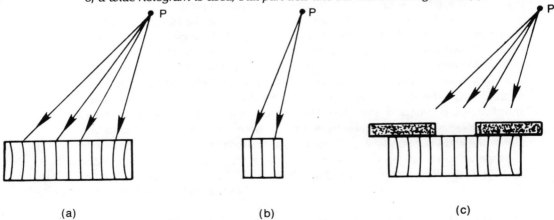

(a)   (b)   (c)

---

[6]"Pictures" of individual atoms have been made holographically using X-rays, but by a special method. For more about this, see Chapter 11.

range of directions than in (b). On reconstruction of the virtual (or real) image with the wide hologram, the viewer can look at P from a range of directions (as though looking through a wide window). With the narrow hologram, the object at P will be seen three-dimensionally again, but you, the viewer, will have little freedom in moving your eye to see it.

If a hologram is broken into pieces, each piece will reconstruct the scene. *Redundancy* is the term applied to this. A less destructive way to observe this is to use the usual diverged beam and an opaque screen with a hole in it to block all but a small part of the plate. Still another way is to let the undiverged laser beam (no lens) fall on the plate. Only a small spot will be illuminated and with some effort you can manage to peer through this to see the virtual image. By using a card, you can discern the real image. (In either case, be sure that you are really looking at an image due to the actual laser beam. Accidental reflections from parts of the beam exit port of the laser may be bright enough to illuminate larger regions of the plate and give images.)

Besides the limitation in the range of viewpoints possible with small pieces of a hologram, there is also some degradation of the quality of the image. (Obviously the inherent graininess of the emulsion becomes significant with very small pieces.) The total number of fringes in a hologram, and not just the spatial density of fringes, is important in determining the image quality. Big is generally better than small in this context.

THE HOLOGRAM IS TWO-FACED

Suppose a transmission hologram is set up with a laser beam shining on it. If you turn the plate front-to-back so the side that was facing the laser becomes the side you look through, you will again be able to see the virtual and real images.

Figure 4.1 provides a basis for a brief discussion of this effect. Suppose that besides the plate shown there were another one displaced vertically downward to occupy a position symmetrically located with respect to an imaginary line between A and B. The interference pattern recorded would be like that in the plate shown, but related to the faces of the plate in front-to-back fashion. If the plate that is shown (after processing) were flipped over without moving it otherwise, it would act as a hologram as would the other plate in *its* position.

The case illustrated is a highly symmetrical one and so is rather special, but this discussion of it should convey the basic idea.

DEPTH OF SCENE AND COHERENCE LENGTH

Most transmission holograms provide views of objects that are small and compactly grouped together. They have small *depth of scene*. This is not due to a limitation in the fundamental nature of holography but rather is associated with a certain property of the laser beam. A laser beam (or any other source of light for that matter) has a limited *coherence length*. This limits the depth of scene that can be achieved.

Refer to Figure 5.10 for an explanation of these matters. The waves *a* and *b* are identical with respect to wavelength and amplitude, and they are in step along their lengths. At such places as A, B, and C they would interfere constructively if superimposed. The result would be the wave *c*, which has the same wavelength but greater amplitude or intensity. Now consider the waves *a* and *b* of Figure 5.11. The wave *b* has a slightly shorter wavelength than the wave *a*. At A, the waves both have crests, and they would interfere constructively there. In the region B the waves have fallen somewhat out of step, and at C they are still more out of step. Eventually at X the waves are completely out of step. The condition for constructive interference at A becomes that for destructive interference at X. Although the drawing does not show it, after traveling another distance equal to that from A to X, the waves will be back in step again. The distance from in-step back to in-step again is twice the distance from A to X and is called the *coherence length*.

Any light source that is not perfectly monochromatic has a limited coherence length because there are constituents in its light that fall out of step along the beam. This is true of the popular helium-neon gas laser, which actually emits more than one wavelength. The wavelengths do not differ greatly but their interplay limits the coherence length. The numerical value of the coherence length depends on the length of the tube in the laser and so is not the same for all helium-neon lasers. Typically, it is several tens of centimeters in magnitude.

Figure 5.10. *The waves* a *and* b *are identical and in step along their lengths. They interfere constructively everywhere and their resultant is the wave* c.

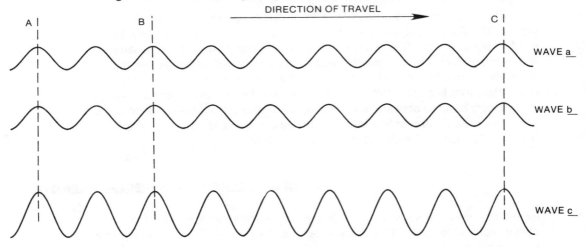

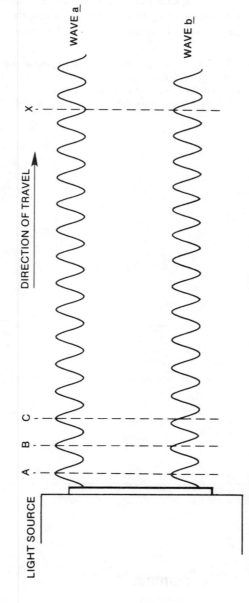

Figure 5.11. Waves a and b differ in wavelength. At A they are in step. Then they become more and more out of step, as at B and C. At X they have become completely out of step.

(For helium-neon lasers with various lengths, the coherence length decreases as the length increases.)

Special lasers incorporate optical devices that suppress some of the wavelengths and can have much greater coherence lengths, possibly extending to many meters. Lasers with coherence lengths of as much as a kilometer (about

Figure 5.12. *(a) Two burst of light are emitted in phase with each other. (b) The bursts have traveled equal distances in the same time and are still in phase. (c) A mirror is used to make the paths unequal, and in a given time interval the burst B lags so far behind A that the bursts cannot interfere at all.*

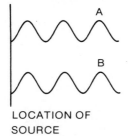

LOCATION OF
SOURCE

(a)

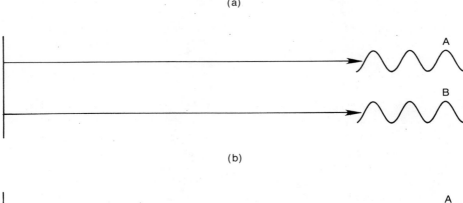

(b)

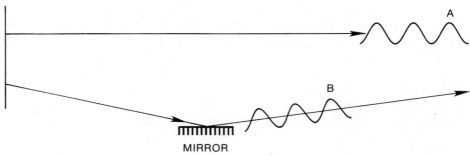

MIRROR

(c)

three-fifths of a mile) exist. Such special lasers are much more expensive than ordinary helium-neon lasers.

In making a hologram the size of the scene is determined by the coherence length of the laser. For example, if the coherence length is one-half of a meter, the depth of the scene must be substantially less than that, lest parts of the scene be not holographed. Thus the practical holographer should know what the coherence length of the laser is. A simple method is given in Chapter 14.

The rest of this section deals with the physical basis for the relationship between coherence length and the making of holograms.

To begin an explanation of this let's refer to Figure 5.12. In (a) two identical short wavetrains start out in phase. Each will be referred to as a "burst." In part (b) the bursts have traveled equal distances to the right in the same interval of time, and they are shown as side by side. If they were superimposed, they would interfere. In part (c) of the figure the case is quite different. The burst A has traveled a certain distance and the burst B has traveled the same distance in the same time, but thanks to the different geometries of the paths, B has not come into position to meet A. In fact, B lags so far behind A that there is no overlap of the two bursts at all, and no interference can take place.

After B travels a sufficient distance further, it will arrive where A was, but in the interim A will have moved on. Whether A and B overlap and can interfere at a point depends upon whether A and B travel paths of the same length from their mutual source to the point in question. If the paths are slightly unequal, there will be partial overlap and weak interference. If the paths are sufficiently unequal, there may be no overlap at all.

The next step is to take the bursts to correspond to samples of light from a laser (or any other source) with limited coherence light. This puts us in a position to understand Figures 5.13 and 5.14. In the first of these the object beams O_s and O_c take paths of different lengths to get from the lens to the holographic plate than does the reference beam R. Depending on what the path differences are, you might get no hologram at all, or one with very low contrast. In Figure 5.14 the configuration has been corrected by using a front surface mirror to lengthen the path of ray R in the reference beam from the lens to the plate. In addition, the objects being holographed have been moved closer together. (The screen is present to block laser light from reaching the plate directly, rather than after reflection from the mirror.)

The golden rule is that the maximum path length difference for the reference and object beams should be made as small as possible.[7] Equalization cannot be brought about for _all_ rays, as you can see by considering the various points such as P, P_1, and P_2 on the plate, and the necessary spread of points in the scene. Hence the rule refers to the _maximum_ path difference.

If the maximum path difference is not zero but is an integral number of

---

[7]A mere piece of string becomes a valuable holographic tool by providing a simple way to compare path lengths.

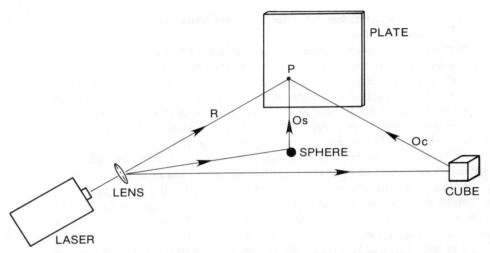

Figure 5.13. *The reference beam is represented by ray R. The object beams from the sphere and the cube are represented by waves O_s and O_c respectively. The three rays travel different total distances from the lens to the point P on the plate.*

Figure 5.14. *The reference beam path length is increased by the front surface mirror, and the sphere and cube have been moved to make all three path lengths more nearly equal.*

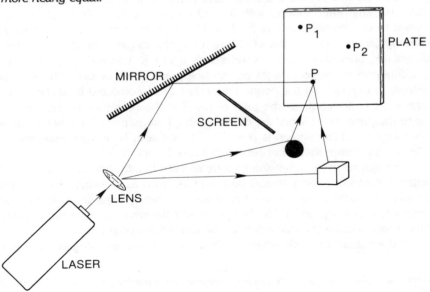

coherence lengths, the condition is suitable for overlap of the "bursts." However, such a setup is of more theoretical than practical interest.

Holograms of very deep scenes have been made, such as of all the things in a room. If you see one, however, you can be sure it was made with a special laser. The coherence length must have been great and the laser probably cost tens of thousands of dollars.

At the other extreme are ordinary light sources. These have some coherence length but may amount to only a fraction of a millimeter. Even if a color filter is used to make the light more nearly monochromatic, the coherence length will still be very short. In this connection it is interesting to know that when Dr. Gabor made the world's first hologram, long before the advent of the laser, he used a mercury lamp with a color filter, but even then had to use a very flat object. (It was a piece of thin transparent material with dark lettering on it.)

MULTIPLE-SCENE HOLOGRAMS

In ordinary photography a double-exposure is usually due to an error and is to be avoided. It results in a confusing superposition of pictures. In contrast, multiple exposure of a plate in holography is often done deliberately. One application of the technique is in an important specialized branch of holography called interferometry or holometry. This subject is discussed in Chapter 11. This section discusses another—and potentially exceedingly important—application known as *multiple-scene* or *multichannel* holography.

Consider the simplest case in which two scenes are recorded in a single transmission hologram. To observe one of the scenes, the hologram is oriented suitably in a laser beam or in color-filtered light from a white light source. To observe the other scene, the hologram is reoriented. In each case only one scene is visible, with no "cross-talk" from the other. For example, in one commercially available two-channel hologram, a model of a motorcycle is seen with one orientation, and a model of a tank after a rotation of the plate.

How such holograms are made and work is easy to understand. With one of the scenes set up, a transmission hologram is made but the emulsion is not processed. While the emulsion is protected from exposure to light, the scene is changed and the plate is reinstalled with a rotation in the setup. The plate is exposed again. The plate is then processed to make the final hologram.

The result is that there are two distinct sets of interference fringes in the hologram. These act on a readout beam of light so differently that there is no overlap in the images. Each image can be seen independently of the other. That this works is evidence of a great difference between holography and ordinary photography.

If two-channel transmission holograms can be made, then how about multichannel recording of greater complexity? How about 500 images in a single hologram so that an entire book could be recorded on one small plate? Given the magnitude of the problem of storing information in business, government, library work and for computing, the question obviously is an important one.

Indeed, very dense recording of information has been accomplished. However, there are complications in this that are not consonant with the present discussion of transmission holograms. For one thing, if the above method for making a two-channel hologram is used for making a many-channel hologram by reducing the amount of the rotation of the plate between exposures, overlap of the images becomes troublesome. This problem can be reduced by making multichannel *reflection* holograms and by other special techniques. This is discussed further in Chapter 12.

COLOR EFFECTS WITH TRANSMISSION AND REFLECTION HOLOGRAMS

The geometric model makes use of only one optical phenomenon—reflection of light from mirrorlike surfaces. Such reflection is not wavelength-dependent. All colors reflect in the same manner. For example, when you look at yourself in a mirror using ordinary room light, you see yourself and your surroundings in natural color. There is no "dispersion," or separation, of the colors. On the other hand, if you pass white light through a prism, the various colors are spread out. The prismatic spectrum is produced not by pure reflections, but rather by another optical phenomenon called refraction.

If the geometric model told the whole story about transmission holograms, then these would be viewable in white light and would give natural color images. In fact the image is more complicated, and usually quite unacceptable. Each object is seen as a smear, sometimes so bad that if you do not know beforehand what the objects are, you may be unable to identify them. Furthermore, there is a spectral distribution of all the color components of the white light across each smear. See Figure 5.1. These color smears require another explanation.

In view of the production of spreadout color spectra by prisms, it would be reasonable to suppose that transmission holograms must have in them some prismlike features. The resemblance is only superficial, however, and no prism model exists. The dispersion of colors that is seen is due to yet another optical phenomenon called *diffraction*. A simple and very successful model that incorporates this phenomenon exists. It is most often called the *zone plate model*. This is discussed in Chapter 9.

Color effects with *reflection* holograms are at least as important and interesting as with transmission holograms. In fact, one of the most important properties of reflection holograms is that they are white-light viewable. They give excellent images when illuminated with small tungsten bulbs or with sunlight, eliminating the need for use of lasers or special monochromatic sources.

To say that they are "white-light viewable" is not to imply that they give either white or natural color images, however.[8] Instead, they give images in single colors, ordinarily the red color of the laser light that was used to make them. They act as their own color filters.

The purely geometric model cannot tell the whole story about such color effects for the same reasons that were stated previously. The model must be supplemented by understanding of another wavelength-dependent optical phenomenon. This is called _Bragg reflection_. This phenomenon and some of the properties of reflection holograms in general are discussed in the next chapter.

---

[8]Nearly true-color transmission holograms have been achieved, but only by a special and difficult technique. See Chapter 11.

Reflection holograms

In viewing the virtual image with a transmission hologram, you look through the plate as through a window, with the light source on the other side. In viewing a reflection hologram, the light source is on your side of the plate, and the situation is more like looking at a mirror. The geometric model alone does not prepare you to understand some striking differences between the two cases. The differences exist in part because an optical effect called *Bragg reflection*[1] is not taken into account by the purely reflective geometric model, though it predominates with reflection holograms and is present to some extent in transmission holograms. Our purpose here is to explain this optical effect and then to apply it to reflection holograms.

SOME GEOMETRICAL OPTICS AND REFLECTION HOLOGRAMS

Let us start by thinking about how the purely geometrical model (aside from Bragg effects) explains the formation of images by a reflection hologram. (See Figure 4.13.) The light reflected from the surfaces in the hologram is perceived by the viewer as if it came from the virtual image. The reflecting surfaces in a reflection hologram are oriented *along* the plate, and this is quite different from

---

[1]This effect is usually called *Bragg diffraction,* but in the present context our name for it is more descriptive of the phenomenon we are discussing.

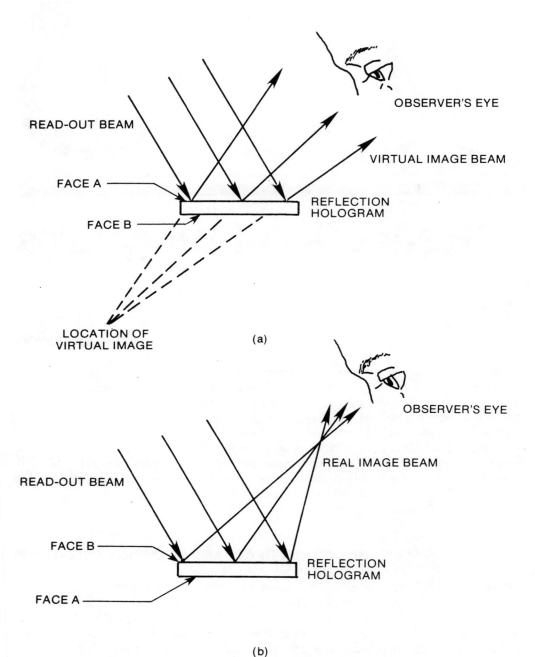

Figure 6.1. *(a) When one side of the reflection hologram is used, the image is virtual and orthoscopic. (b) When the other side is used, the image is real and pseudoscopic.*

the across-the-plate orientation of transmission holograms. However, the basic idea of reflections to produce the virtual images applies in both cases.

The production of _real_ images by transmision holograms and by reflection holograms are quite different. A transmission hologram illuminated as in Figure 5.1 gives a virtual image and a real image simultaneously, which is understandable according to the geometric model. There is also a direct beam straight through the plate from the light source that carries no image information. What happens with a reflection hologram is shown in Figure 6.1. In (a) the viewer holds the hologram in a readout beam of light to see the virtual image. In this _particular_ arrangement there is no real image.

However, there _is_ an arrangement which gives a real image. To see it, the reflection hologram must be flipped over so that the faces A and B are interchanged, as shown in (b). This real image is pseudoscopic, while the virtual image seen as in (a) is orthoscopic.

Only the effects are described here; their explanations are given later. But if you have on-hand a suitable reflection hologram, you can easily make observations that are pertinent. The hologram must be viewable from both sides without a permanently attached opaque backing. By viewing it from one side and then the other you can see the images as in (a) and (b) in Figure 6.1, and their orthoscopic and pseudoscopic natures. This is more striking with some holograms than with others, and it is best if the object has substantial depth. We have shown the effect to many people, using holograms of statuettes, ancient coins, and so on, and have found that some people react strongly to the curious contrast between the two images. Evidently some degree of sophistication is required for the reaction to occur. In any case, it is something that any student of holography should experience.

AN OVERVIEW OF THE NATURE OF BRAGG REFLECTION

In an actual reflection hologram, the partially reflecting surfaces will form a complex array in the plate, even for a relatively simple holographed object. As you start to consider the optical effects that take place in reflection holograms, it is best to think of an idealized case in which the surfaces are plane, parallel, and uniformly spaced. Figure 6.2 shows such a stack of surfaces, with a beam of monochromatic coherent light falling obliquely on the top of the stack. When the light encounters the top surface, some of it is reflected back upward and to the right, and some passes through to the next surface. At this second surface some more of the light is reflected and some is transmitted toward the third surface. Similar reflections and transmissions occur at subsequent surfaces, though with decreasing amplitudes due to losses in the light.[2]

---

[2]For simplicity, we ignore multiple reflections of light back and forth in the region between any two adjacent surfaces and effects of refraction.

INCIDENT LIGHT

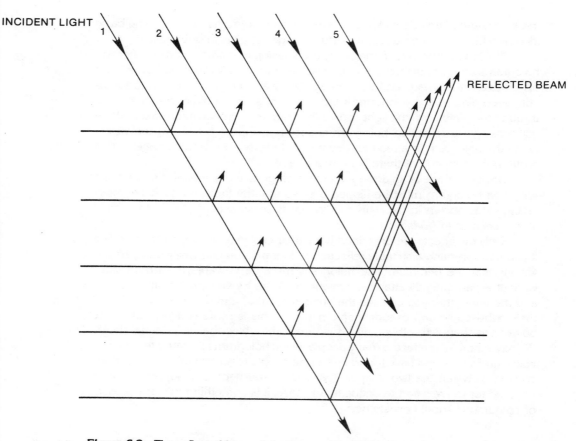

REFLECTED BEAM

Figure 6.2. *The reflected beam from the stacked surfaces is the result of light reflected from the various surfaces and interfering with each other as they emerge from the stack.*

By now you know that if *two* coherent light beams are superimposed, there will be interference effects. If *many* coherent beams are superimposed, there will also be interference. You can correctly anticipate that the interference of the reflected beams will be important. This combination of multiple-surface reflections and the consequent interference is Bragg reflection.

BRAGG REFLECTION IN SOME SPECIAL CASES

Let's take up first the case of just two partially reflecting planes with the light falling perpendicularly on the top surface as in Figure 6.3. In (a) some of the incident ray *a* reflects directly as ray *b*, and some goes on as ray *c*. In (b) of

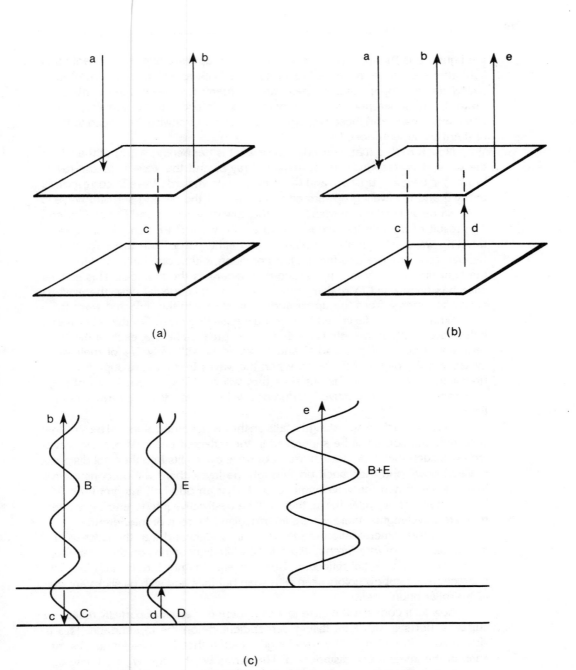

Figure 6.3. With normal (perpendicular) incidence on two parallel surfaces, constructive interference in the reflected light takes place when the wavelength is twice as great as the distance between the surfaces. For interpretation of the parts (a), (b), and (c) of the figure, see the accompanying text.

77

the figure, *d* is that part of ray *c* which reflects upward, and *e* represents this light after it emerges from the upper surface. With an extended beam of light falling on the upper surface, there will be many rays like *b* and *e* traveling upward, and superimpositions will occur. For simplicity in the drawing only *b* and *e* are shown, and these are shown as displaced sidewise. You need to think of them as superimposed.

Now turn to a *wave* description of what is happening. In (c) of the figure the wave B corresponds to the reflected ray *b*, and the wave C to the ray c. Wave B is traveling upward, and C is traveling downward. Wave D corresponds to ray *d* and is traveling upward after reflection at the second surface. Wave E is the same as D after emergence. It is the upward waves B and E that interfere. The result of the interference is shown as the wave B + E. This is an upward traveling wave with greater amplitude than either B or E. In order for this to happen (constructive interference), a condition is that one-half wavelength of the light fits precisely into the space that separates the surfaces. This special case has been used in making the drawing. In this special case the wave B arrives to emerge from the upper surface in step with the reflected wave E.

What we have described is a strongly reflecting layer. On the other hand, if the wavelength of the light is made twice as great as before, each of the wave segments that correspond to C and D would fit with only 1/4 of their wave cycles into the depth of the layer. When the waves E and B are superimposed (traveling upward above the surface), they will be 1/2 wavelength out of step. The condition for destructive interference will be met. Now we have a nonreflective layer.

If light of *various* wavelengths falls on the layer, some colors will be strongly reflected, and some will be suppresed in the reflected beam. When the wavelength is such that an integral number of wave cycles fits into the total distance traveled down and then back up through the layer, there will be constructive interference. When the wavelength is such that an odd-half integral (1/2, 3/2, 5/2 . . .) number of cycles fits in, there will be destructive interference. For intermediate wavelengths, there will be intermediate degrees of interference.

With this generalization made, you can understand that the layer will be a *color selector* or *interference filter,* for if white light shines on the layer, there will be color-preferential reflection. Such an effect is not rare in daily life. The characteristic rainbow colors seen with soap bubbles and with oil slicks are due to a similar phenomenon.

Now let's consider the case in which there is a stack of partially reflecting surfaces, but still with light falling perpendicularly on the top surface. This is illustrated in Figure 6.4. The wavelength is such that 1/2 wavelength fits into each of the intersurface distances d. The waves W_1 through W_5 are those that emerge traveling upward after reflections in the stack. As they emerge, their various amplitudes are smaller and smaller, according to their depths of penetration and consequent losses. But they are in step, and as a result there is

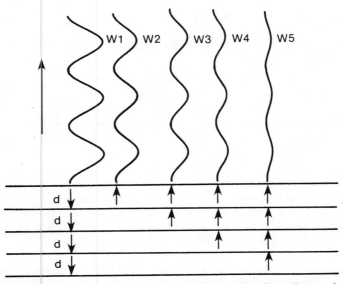

Figure 6.4. The waves W_1 through W_5 emerge from the stack in step with each other after reflections at various depths, if the wavelength is two times the distance between adjacent surfaces.

constructive interference. Because the interference involves a number of cooperating beams of light, the reflection will be strong.

In the opposite case of destructive interference—for some other wavelength, of course—there will be very effective mutual cancellations. The multiple-layer system will be a highly selective color filter.

OBLIQUE INCIDENCE OF LIGHT

Figure 6.5 shows an extended beam of monochromatic light falling obliquely on stacked surfaces. (For simplicity only two surfaces are shown.) A and A' are two rays in the incident beam. Some of A reflects immediately as B, and some of A' as B'. Some of A passes through the upper surface, travels on as C, reflects as D from the lower surface, and finally emerges as E. B' and E are superimposed as they leave the stack traveling upward, and they will interfere.

Whether the rays B' and E interfere constructively, destructively, or in some intermediate way, depends as before on the wavelength and on the intersurface distance, but now it also depends on the angle of incidence of the light. For some angle of incidence there may be strong reflection, with B' and E in step under the given conditions. For some other angle of incidence, with all the other conditions remaining the same, there may be no reflection, and

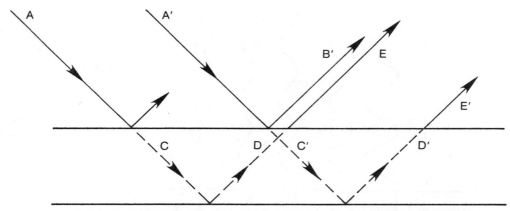

Figure 6.5. *With oblique incidence of the light, the rays B' and E (reflected from the first and second surfaces, respectively) interfere in a manner that depends on the wavelength, on the surface separation, and on the angle of incidence.*

B' and E may be completely out of step. If B' and E are in some intermediate phase relationship between the two extremes, there will be some intermediate intensity of reflection.

If *white light* is used, there will be different angles for maximum intensity of the reflected light for the various color components in the illuminating beam. For example, the conditions shown in Figure 6.4 may be right for strong reflection of light of one particular spectral color, which the Bragg effect selects from the full range of colors in the incident light. If the angles are changed, the Bragg effect will select another color to predominate in the reflected beam.

This is the manner in which color effects seen with soap bubbles and oil slicks can be explained. In an introductory physics textbook such things are explored in detail. We will not have to do that here because holograms do not meet the usual textbook-case conditions. In holograms the stacked surfaces are not flat, parallel, uniformly spaced, all equally reflecting, and in fact are so granular on a microscopic scale that they are not even continuous surfaces.

THE BRAGG EFFECT WITH REFLECTION HOLOGRAMS

Figure 6.6 shows a model of a reflection hologram. Sheets of ordinary 8½-by-11-inch paper are spaced vertically with a separation of about 1/2 inch between adjacent sheets. These represent the interference fringes across the depth of a typical holographic emulsion. Among the most commonly used emulsions are Kodak SO-173 and Agfa-Gevaert 8E56 and 8E57. These are about 6 millionth's of a meter thick. The separation between fringes is about 1/2 wavelength in typical cases. For the usual red laser light this separation is about 1/4 millionth of a meter. Hence the model is fairly realistic vertically.

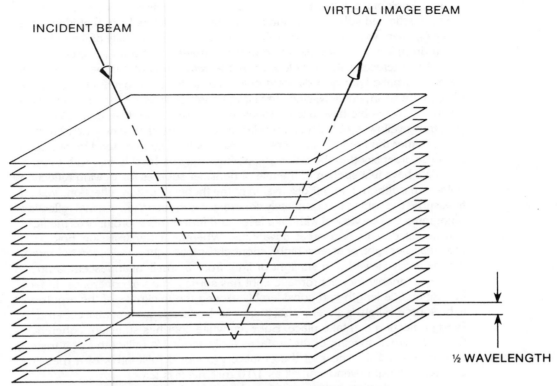

INCIDENT BEAM

VIRTUAL IMAGE BEAM

½ WAVELENGTH

Figure 6.6. The paper sheet model of a typical reflection hologram.

With these dimensions in mind, let us consider what happens when light falls on a hologram of this kind. In Figure 6.6 only one ray is shown—the one that penetrates the full thickness of the plate before reflecting. In the full situation, light is partially reflected and partially transmitted from each of the surfaces, and the large number of surfaces favors strong Bragg reflection.

Suppose the hologram was formed and illuminated with red laser light. In the idealized case we discussed earlier strong Bragg reflection will occur for one precise angle. With an actual hologram it takes place over a wide range of angles. In a typical case, you may be able to see the image even through a range of angles as large as 20 or 30 degrees.

This difference between what really happens and what you expect on the basis of the simple model is attributed to the combined effects of the thickness of the emulsion, the density of the surfaces, the reflectivities of the surfaces, and the nonlinear nature of the recording medium.[3] The theory shows that with

[3]A more detailed (but mathematical) discussion of this can be found in *Optical Information Processing and Holography* by W. Thomas Cathey (New York: John Wiley & Sons, 1974) pages 150–175.

an actual hologram, if the angle is that angle A_B predicted in the idealized case, then the reflected light is most intense. If the angle differs from that ideal-case angle A_B, there will still be Bragg reflection, but with the intensity of the reflected beam dropping off as the deviation of the actual angle from A_B increases.

It is interesting that a hologram made with red light can be viewed with monochromatic light of a different color. To do this, it is necessary to tilt the plate in the beam suitably, since the Bragg angle is different for different wavelengths. Few people have access to lasers that emit light of various colors, but the strong yellow light from sodium lamps can be used to view a hologram in a color other than red. Color filters and ordinary light sources tend to give such low initial intensity that the color-angle effect may be hard to see with them.

We have just spoken of the use of nonlaser sources for viewing reflection holograms, although in our discussion of the theory of Bragg reflection, we had to assume the light to be completely coherent. It is indeed the case that reflection holograms can be viewed without a laser, and furthermore, ordinary *white light* can be used. This is a great advantage. Whereas a transmission hologram requires laser light, or at least highly monochromatic light, a reflection hologram can be easily viewed with sunlight or light from a small incandescent lamp.

Reasonably monochromatic light has some temporal coherence. In fact, the coherence length can considerably exceed the thickness of the emulsion. If the source of light is small in size, there will also be adequate spatial coherence, Bragg reflection can take place. But even white light has sufficient coherence. Given the small spacing of the surfaces (only about 1/4 millionth's of a meter) and the limited thickness of the emulsion (about 6 millionth's of a meter) a coherence length even as small as 1/10 millimeter is large!

Thus reflection holograms are white-light viewable, which makes them easy to view and exhibit in holography shows, and which makes feasible their use as desk ornaments, pendants, etc.

COLOR EFFECTS

"White-light viewable" does not mean "viewable in full color."[4] If a hologram was made with red laser light and then illuminated with white light, the image would be seen in red. The hologram acts like a red-selecting color filter. The red image can be seen over a range of angles that may deviate considerably from that predicted by the simple theory.

Actually, the hologram also is a color filter for other colors, but ordinarily you see the image only in red. The reason is twofold. First, the intensity of the Bragg-reflected light is much less for colors other than red, and second, the range of angles over which the red light is reflected overlaps the Bragg angles for other colors. Other colors are submerged under the red reflected light.

[4]See Chapter 11 for a discussion of true-color.

A reflection hologram does not produce *only* a red image with white light illumination. Some can be moved about in the beam until eventually an image is seen quite distinctly in some other color. For example, there may be a main image of a bright red rose, and a secondary image of the same rose in a blue-green color. For this to happen, the Bragg angle for this other color must be so greatly different from the red light that at most only a small residual red reflection occurs. This leaves the other color visible.

A reflection hologram viewed with white light often gives a greenish image rather than a red one. This effect is quite different from the color effects just mentioned. It is due to shrinkage of the emulsion during the chemical processing. This shrinkage reduces the spacing between the Bragg planes, so that the hologram appears to have been made with light of a shorter wavelength than was actually used. The nature of the emulsion medium and the processing usually shift the effective wavelength to the green spectral region. The image in such cases has a greenish tint. Even with shrinkage, the hologram can still be viewed with the original red laser light, but the central Bragg angle will be different from how it would be if there had been no shrinkage.

MORE ABOUT IMAGES WITH REFLECTION HOLOGRAMS

Early in this chapter a virtual orthoscopic image and a real pseudoscopic image were discussed, and how they can be seen with a reflection hologram. Here we will discuss briefly the production of these images. Figure 6.7 shows a simplified conception of the hologram, with the partially reflecting surfaces curved and essentially parallel. Part (a) shows the usual configuration for exposing the emulsion, with the film relatively near the object, and the general shape of the interference fringes and how they lie in the emulsion. As (b) stresses, the finished hologram is different as seen from the side of face A from how it is seen from the side of face B. The plate "remembers" the orientation it had when it was made.

If the plate is reinstalled in its position in (a), with the illuminating beam coming from the left and with the original object removed, the light incident on the plate will encounter surfaces that are convex toward the source. The reflected rays may then diverge as they move back toward the left, and this would correspond to production of the virtual image. But this cannot be the whole story, for the reflection hologram does not simultaneously produce a real image, although a transmission hologram does (in which the readout beam may also fall on convex surfaces). The Bragg effect must be invoked for a fuller explanation.

Figure 6.5 shows how a stack of flat and parallel surfaces can produce a strong Bragg-reflected beam (for a given wavelength and surface spacing) at only one angle. If in looking at the figure you imagine the surfaces to become

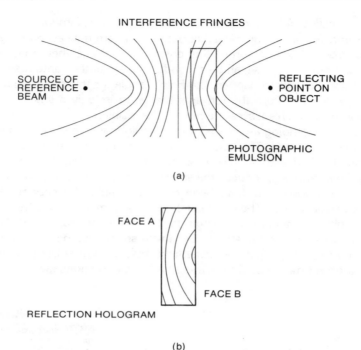

INTERFERENCE FRINGES

SOURCE OF
REFERENCE •
BEAM

REFLECTING
• POINT ON
OBJECT

PHOTOGRAPHIC
EMULSION

(a)

FACE A

FACE B

REFLECTION HOLOGRAM

(b)

Figure 6.7. *(a) Schematic representation of a setup for making a reflection hologram. (b) The finished plate contains surfaces that are different when viewed from the side of face A and from the side of face B.*

curved gently, convex upward, you can understand why the reflected rays will diverge, and accept that the condition for strong Bragg reflection still cannot be met by any other rays (such as would be required for a set of convergent rays).

If instead you imagine the surfaces in Figure 6.5 to become curved gently, and concave upward, you can see how the reflected rays will be convergent and why only these real-image rays undergo strong Bragg reflection.

Now let's try a simple experiment that shows another interesting property of some reflection holograms. All you need is a reflection hologram that does not have some opaque backing on one side, so that you can shine light through it. The experiment consists of using the reflection hologram as if it were a transmission hologram. That is, you face the source of white light or the laser (with a spreadout beam), interpose the hologram between your eyes and the source, and look through the film. You should be able to see the virtual image. It will be fainter than the image you see when you use the plate normally as a reflection hologram, because strong Bragg reflection still takes place, and *that* back-reflected light is not available to your eye to see the virtual image. If the hologram gives only a somewhat bright image in the reflection mode, the transmission mode image may be quite bright, but if your hologram produces a very brilliant reflection mode image, you may see only a very faint and indistinct transmission mode image.

The image will be seen on the other side of the plate, but probably not very far over there because reflection holograms are usually made with the object near the plate. If you use a white-light source to view the transmission mode image, you will see it in a color other than red. We will call the color a bluish ashen color. It is the result of the reflection of much of the red component from the white light from the source, together with some of the green (due to shrinkage of the emulsion). What colors are reflected back are subtracted from the light available for viewing the image. If your hologram gives you a viewable virtual image when used as a transmission hologram, you probably will also be able to see a real image. It should be pseudoscopic, fainter than and colored differently from the pseudoscopic reflection mode image.

This use of a reflection hologram as a transmission hologram is an instance in which properties of both kinds are present in one plate. (More about this in Chapter 7.)

A SPECIAL FEATURE OF NEAR-SCENE REFLECTION HOLOGRAMS

A transmission hologram contains optical information about _each_ point on an object in the scene spread out over the _whole_ hologram. This accounts for the property of redundancy. But a reflection hologram made with the object close to the emulsion can form the interference fringes that pertain to any one point on the object only in a small region that is near to the object point. The optical information about the point is located in nearly one-to-one correspondence with the object points.

In simpler language, the result is that light shone onto the plate to view the image is reflected from the plate rather as it would be if it were shone onto the original object—point by point—whereas the whole plate contributes to the image of each object point in the case of a transmission hologram.

There is another kind of hologram that also has this property. It is called a _projection_ or _stand-out_ hologram, and it is discussed in Chapter 10.

Phase holograms and some other types

Transmission and reflection holograms are two contrasting types. In the world of holography there are still other pairs of contrasting kinds: surface and volume, some are bleached and unbleached, thick and thin. Such distinctions and their significance are discussed in this chapter.

PHASE HOLOGRAMS AND BLEACHING

In ordinary photography light falling on the emulsion activates the silver halide grains, strongly where the light is intense and less so elsewhere. When the negative is made there are regions that are quite dark (those exposed to high light intensities) and less dark regions, shading down to virtually transparent ones. In holography the situation is similar, yet the entire hologram appears dark. The reason is that the interference fringes are very closely spaced and extend over the whole film.

Obviously such a hologram cannot transmit light very well. It is appropriately called an *absorption hologram*. The *efficiency* of a hologram is defined as the fraction of the incident light that is incident on it that is actually used in reconstructing the images. For a transmission hologram with a typical thickness

87

of several microns, the efficiency may be only 2 or 3 percent! For reflection holograms the efficiency may be only 5 to 7 percent.

The situation is much better with *phase holograms*. Phase holograms of both the transmission and reflection types can have efficiencies that approach 100 percent. With so much of the darkness that *absorbs* light absent, a phase hologram can put more of the available light into the image, which will be more brilliant. The improvement is so great that it is little wonder that most finished holograms today are phase holograms.

Some phase holograms are made with holographic media of special kinds that do not blacken. An example is a kind called *dichromated gelatin*. When this substance is exposed to light and processed by special means, the optical information in the form of interference fringes is recorded, but by means of variations in the "index of refraction" throughout the medium, which remains transparent. The following section will discuss this effect further. Another example of a special phase hologram medium is *thermoplastic*. This reacts to the heat energy carried by light incident on it. Still other phase media are known.

Phase holograms are usually created by converting an ordinary absorption type of hologram made with ordinary silver halide emulsion into a phase hologram by a step in the processing of the film. After the film has been developed, but before the processing has been completed, the film is soaked in a chemical solution to bleach it. The bleaching converts the dark grains in the negative into other silver salts that are transparent. The bleaching does not erase the interference fringes but rather transforms them.

BLEACHED HOLOGRAMS AND THE GEOMETRIC MODEL

Using silvered mirrors is part of daily life, so you probably understand the reflection of light from opaque reflecting surfaces. Because of this, the idea of reflection from the dark hyperbolic surfaces in a hologram is easily accepted. But how can the reflections be understood in the case of a bleached hologram?

Reflection from clear surfaces actually is also common in daily experience. For example, you can of course see yourself reflected in a store window, often with very high efficiency. Again, you can see that there are bubbles or other defects in a sheet of glass, although they are not opaque.

Technically such reflections are said to take place because the light encounters regions where the "index of refraction" changes. In Figure 7.1 there is an interface between two media, which might be air and glass, or a clear oil on water, for examples. Let's assume that the media are both solid, since this corresponds to transparent fringes in the body of an emulsion. The speed of light generally varies according to different media, and is less than in vacuum or in air. For example, the speed in one kind of glass may be 2/3 as great as in vacuum. For another, the speed in a plastic may be 3/5 as great as in vacuum. In any case, the *index of refraction* for a medium is defined as the ratio of the

RAY OF LIGHT

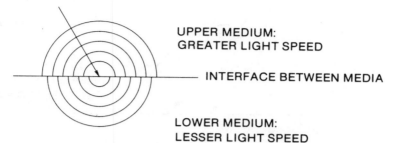

UPPER MEDIUM:
GREATER LIGHT SPEED

INTERFACE BETWEEN MEDIA

LOWER MEDIUM:
LESSER LIGHT SPEED

Figure 7.1. Light falling on the interface between two optically different media generates Huygens' wavelets, which spread out at different speeds upward and downward.

speed of light in vacuum to the speed in the medium. In the two examples given, the index is 1.50 in the first case and 1.67 in the second.

It follows that if light travels initially in one medium and then encounters an interface with another medium that has a different index of refraction, the speed of the light will change when it crosses the interface. In Figure 7.1 a ray of light does just that. According to Huygens, the point on the interface where the ray arrives becomes a source of spherical wavelets. Some of these spread upward from the point, and some spread downward. Because of greater light speed in the upper medium in the case illustrated, each wavelet will travel farther in a given time interval than will the corresponding wavelet in the lower medium.

In Figure 7.2 this is applied to the case in which initially there is an extended beam, rather than a single ray. The various wavelets traveling up from the interface have an envelope, and this is the *reflected* wavefront. The wavelets traveling in the lower medium also have an envelope, and this is the *refracted* wavefront. More detailed analysis shows that the reflected beam forms an angle at the interface that is equal to the angle formed by the incident beam, just as is true in reflection from a silvered surface such as an ordinary mirror. The refracted beam changes its direction from that of the incident beam.

In this simple way you can perceive the physical basis for saying that light undergoes reflection and refraction when there is a change in the index of refraction. In a bleached hologram, or in any phase hologram, it is the great reduction in loss of light through absorption that is important, and the optical information in the interference fringes remains. This information is of two kinds: phase and amplitude information. (See Chapter 4.) The amplitude information is recorded by *variation in the index* in the medium, for these variations affect the reflectivity. This is in lieu of variations in reflectivity because of varying degrees of darkness in the partially reflecting surfaces in absorption holograms. Phase information is recorded in the *shapes of the regions,* as in absorption holograms, because the shapes depend on phase relations between the reference beam and the object beam when the hologram is made.

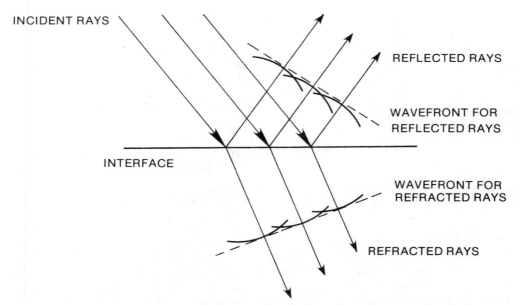

Figure 7.2. *Some of the incident light is reflected and some is refracted. The angle of reflection is equal to the angle of incidence. The angle of refraction is different, and such that the refracted beam "bends" toward the normal to the interface.*

SURFACE HOLOGRAMS

In the discussion above it was taken for granted that the interference fringes are recorded throughout the volume of the hologram. Actually, holographic information can also be recorded on the *surface* of a medium. An important case is a technique in acoustic holography in which the fringes appear as ripples on the surface of a body of water. (See Chapter 11.) In another method, an oil film is scanned by a modulated electron beam, producing deformations of the surface of the film. In still another method, thermoplastic is used.[1]

You don't need to resort to such exotic techniques to get a surface hologram. In fact, when a hologram is being made in the usual way with an ordinary silver halide emulsion, during the fixing process residual undeveloped grains are removed from the emulsion, and there is partial collapse of the substance, greater and lesser in various places. As a consequence of this interior effect, the overall thickness of the emulsion changes from region to region, and the

---

[1]Details may be found in *Optical Information Processing and Holography* by W. T. Cathey (New York: John Wiley & Sons, 1974, pages 141 ff). This is also a good source of information about the differences between surface and volume holograms in general.

surface takes on a texture that can be described roughly as rippled. The details of the ripples constitute a surface hologram.

If the surface is coated very thinly with reflecting material to retain the surface texture, the image can still be viewed in reflected light. It may even be brighter than before, but because a surface cannot contain the wealth of optical information that a volume can, such an image can be expected to be of inferior quality.

Notice that the bleaching of a holographic emulsion can also change the emulsion thickness, and produce the effect just described. An interesting consequence of this is that it makes possible the mass production of copies of holograms. After the hologram is duplicated, the surface is plated with metal. Then the metallized surface is used to make imprints into plastic, much as a master phonograph disc is used to stamp out numerous copies.

THICK AND THIN HOLOGRA

A thick hologram can be defined by considering a reflection hologram. The hologram is thick if the emulsion depth is great enough to contain a substantial number of Bragg-reflecting surfaces. If it can contain only a very few surfaces, the hologram is thin. This is not a very precise definition, but an emulsion roughly ten microns deep, which would contain some tens of surfaces (assuming red laser light), is considered thick. An emulsion only about one micron deep, which would contain only one or two surfaces, is considered thin. Because a strong Bragg effect is desirable for a _reflection_ hologram, a reflection hologram should be thick.

For _transmission_ holograms the distinction between thick and thin has to do with the sizes of the reflecting surfaces that are present between the faces of the hologram. This is illustrated in Figure 7.3. In (a) a thick transmission hologram is shown. In (b) the contrasting case of a thin one is shown. In the ultimate case of an extremely thin emulsion, only striations that lie on a surface can be present. This would be considered a surface hologram.

Many of the properties of thick transmission and reflection holograms can be understood using the geometric model plus Bragg reflection. Yet the geometric model is not capable of explaining thin holograms. As Dr. Jeong says:

> As is true with all models, our geometric model for holography must break down at a limit. This limit is the "thin" hologram. A hologram is considered thin when the separation between the hyperboloidal surfaces exceeds the thickness of the emulsion. At this point our model will start giving the wrong answers. For example, a thin reflection hologram according to the model should give a white image when illuminated with white light, while in reality no image is formed in this way. Also, since there is no longer any Bragg effect, a transmission hologram should also be given a white image if white light

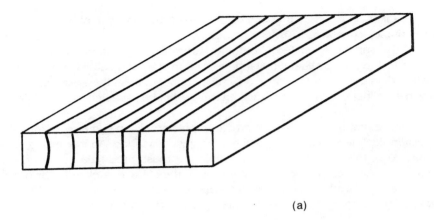

(a)

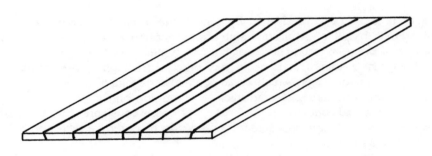

(b)

Figure 7.3. *(a) A thick transmission hologram has substantial portions of the hyperbolic surfaces in its volume. (b) A thin transmission hologram essentially has only striations that lie on its surface.*

is used for the reconstruction. The fact is that the image will be smeared in a continuous spectrum.[2]

You may be thinking that extremely thin holograms are of much less general interest than are thick holograms. This is true, so the breakdown of the geometric model in these special cases is not ordinarily a matter of concern. That is not the whole story, however, for some properties of thick holograms also elude the explanatory powers of the geometric model. These properties, and those of thin holograms, will be dealt with in Chapter 9.

[2]T. H. Jeong, "Geometric Model for Holography," *American Journal of Physics,* August 1975, page 717.

The distinction between a transmission and a reflection hologram is usually very sharp. In the first case the sources of the reference beam and the object beam lie on the same side of the plate, as in Figure 4.1. For a reflection hologram, the configuration is more as in Figure 4.12, with the plate essentially between the light source and the object. These or even more distinct configurations are often used and result in holograms easily classified as one kind or the other. Yet often an arrangement is used in which the locations of the source, object, and plate and the orientation of the plate make the matter of classification less clear. As already noted, a hologram may be one kind and yet show properties of another. (See Chapter 6.)

Figure 7.4 shows some sample configurations out of the infinite varieties possible. In (a) point source S and small reflective object O are used to make a transmission hologram. Because S and O are very close together, the hologram will be nearly *in-line.* This means that when the image O' is reconstructed as in part (b), you have to look almost back into the laser beam to see the image. In this case the hyperbolic surfaces are essentially perpendicular to the faces of the plate, and the Bragg effect is virtually absent.[3]

In (c) source S has been moved so that the reference beam comes to the plate more obliquely, and the surfaces are tilted in the plate. When this off-axis hologram is reconstructed, a ray in the reconstructing beam can encounter a number of surfaces along its path, as shown in (d). Because of this, the hologram could be called a Bragg-effect transmission hologram.

In (e) there is a clear configuration for a reflection hologram. The surfaces now lie nearly parallel to the faces of the plate. This is favorable for Bragg reflection during readout of the image. In (f) the conditions for Bragg reflection have been optimized, though at the cost of in-line viewing. To avoid this, a configuration between those in (e) and (f) would be used, but more like the latter, while Bragg reflection would still be predominant.

The preceding discussions considered cases in which the properties of one kind of hologram could be detected with a hologram nominally of the other kind. Let's conclude this section with an example of how these concepts can be useful. Consider the following statement: "The importance of intermodulation noise is not great with white light reflection holograms." Once the meaning of *intermodulation noise* is understood, this becomes readily decipherable. Figure 7.5 shows the making of a reflection hologram of an object. Source S and the object are on opposite sides of the plate, and such surfaces as the one labeled

---

[3]Sometimes such a hologram is said to be *plane* because the volume of the emulsion is not being used to the full. Other types would then be called *volume* holograms. Unfortunately, the technical terminology in holography has not been universally agreed on, so you must watch out for usages that differ from what you might expect.

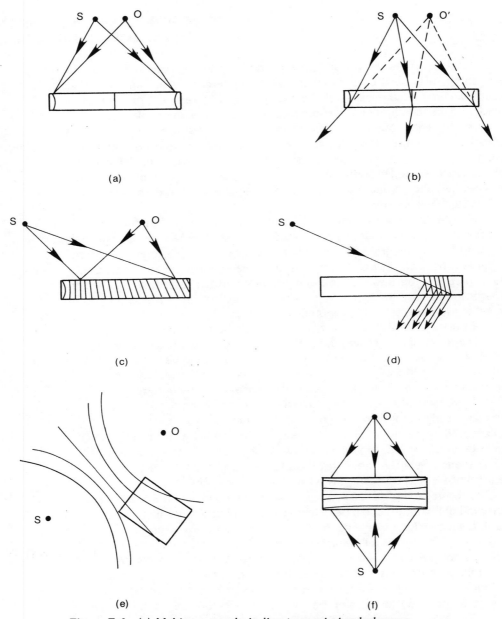

Figure 7.4. (a) Making a nearly in-line transmission hologram.
(b) Reconstruction of the virtual image. (c) An off-axis configuration. (d) How
Bragg reflection can occur to some degree with a transmission hologram.
(e) Making a reflection hologram with an off-axis arrangement. (f) The in-line
reflection hologram configuration.

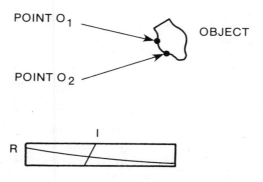

Figure 7.5. *The interference surface R is the kind wanted in a reflection hologram. The surface I (formed by interference between beams from two points 0_1 and O_2 on the object) is unwanted.*

R lie along the length of the plate. But rays from any two points such as O_1 and O_2 on the object will interfere in the plate and produce interference fringes. These fringes are unwanted, and if their consequences are visible in the reconstructed image, they produce "intermodulation noise." However, the fringes produced by O_1 and O_2 will be oriented in the plate much as in the case of the one labeled I. Such fringes as I will constitute a transmission hologram, and with fringes such as R a reflection hologram. If the hologram is viewed as a reflection hologram, with a readout source at point S, the fringes of the kind represented by I will throw the light in directions far from that used to view the image, and the intermodulation noise will not be troublesome.

SOME MORE TERMINOLOGY

In the literature of holography you will hear of *Fresnel* and *Fraunhofer* holograms. A Fresnel hologram is produced when the object being holographed is near the plate, so that the object beam wavefronts are spherical. A Fraunhofer hologram is produced if the object-to-plate distance is so great that the object beam wavefronts are virtually plane. The details of the interference pattern recorded in the plate will be different in the two cases for a given reference beam. Our purpose here is only to say enough about the phraseology to keep you from being baffled when you encounter the terms. Further discussion will be found in Chapter 9.

Now let's briefly consider what is meant by a *Fourier* hologram. Consider a small light bulb. Light from the bulb that falls on a screen is unfocused light, and an image is not formed. If a converging lens is placed between the bulb and the screen and the distances are adjusted suitably, a focused image can be formed. Once this condition is met, if the screen is moved closer to or farther from the lens, there will be only a defocused blur. Yet optical information about the source is present in the light beam. It is said to be in the form of a Fourier transform. If the beam is now allowed to fall on an emulsion, and a coherent reference beam is also shone on the emulsion, an interference pattern will be formed. The developed plate will be a Fourier hologram. It cannot be viewed merely by shining the original reference beam on it, but a lens can be used to reconstruct the image. This kind of holography is an exception to the common characterization of holography as lensless photography. Fuller discussion and an application are given in Chapter 11.

The phrase *lensless Fourier hologram* refers to a different use of the word "Fourier." This is a transmission hologram made with the source of the reference beam and the object being holographed in the same plane. Such a hologram has only some properties of a true Fourier hologram, so the title is a misnomer.

Some basic setups
for making holograms

This chapter describes a number of ways to make transmission and reflection holograms. If you want actually to make holograms you will of course need practical how-to-do-it information as well. Chapter 13 discusses the material requirements in much greater detail than here. Chapter 14 takes up the making of holograms in the style of a laboratory manual.

THE ESSENTIAL APPARATUS AND FACILITIES

You don't need a lot to make holograms. The permanent items are a suitable laser; a table that is shakeproof to holographic standards; a few simple optical components (small lenses, front surface mirrors, and a beam splitter or two); some mounts to hold the optical components in place on the table; and a few very simple ancillary things, such as a block of wood to serve as a shutter, pieces of cardboard to screen off unwanted light, and trays to hold processing chemicals. The expendable items are photographic film or plates suitable for holography and chemicals for developing and bleaching the emulsions. Finally, the table has to be housed in an area that can be darkened when a hologram is being made, and you need an area that can be darkened during processing of the film.

In the rest of this section we will stress the viewpoint of an experimenter who is concerned with simplicity and low cost. The same requirements can be met at a total cost that would be very large by any reasonable standards. Gemlike optical components, precision mounts, and the like can be purchased, but they can be very expensive. The essential laser can cost many thousands of dollars. At the same time, light is indifferent to what the experimenter *paid* for the apparatus, and inexpensive items also function very well. Better results are not guaranteed by greater expenditure of money.

You can set up a holographic facility and make good holograms at about the cost of setting up an ordinary photographic facility at a modest level.

The Table

When you make a hologram you are recording in photographic emulsion a very complicated interference pattern, with details as small as a wavelength of light or smaller. These details must remain stationary during exposure, or no hologram will result. In order for the fringes to be captured, there must be no vibrations or other movements of the table—or of apparatus on the table even if the table is motionless. All must be immune to footfalls in the room, trucks passing on a nearby road, air currents, and the like.

It may seem to be a formidable challenge to achieve such stability, but actually it can be done by anyone, simply and cheaply. The simplest design is shown in Figure 13.1. A wooden slab rests on three or four partially inflated inner tubes or bicycle tires, and these in turn rest on a sturdy table or bench. The slab can be two or three feet wide and perhaps five feet long. Some 3/4-inch plywood will serve very well. The inner tubes provide a flotation system that helps protect the slab from disturbances and that helps dampen out movements quickly when they do occur.

When the table has been set up, you must test it. This is easily done by a simple optical method, using only a few pieces of apparatus. The procedure is itself very interesting and is described fully in Chapter 14.

Lab and Darkroom Facilities

The table must be housed in a room large enough to hold it and yet allow room for working around it, and that can be darkened so that extraneous light will not fall on the emulsion. A concrete floor, as in a garage or basement, is better than a wooden floor. Electrical power is necessary.

For processing exposed film you need no more than enough table space to hold four or five small trays in an area that can be darkened. The trays will hold chemical solutions and water rinses. There must be water and sink available, but these need not be in the same room, since the trays can be carried elsewhere for filling and emptying. It is convenient to have this modest darkroom in the same room as the holographic table, but again the exposed film can be carried (in a light-tight box or sack) elsewhere for processing.

The Laser

It is not unreasonable to think that in the future we may have inexpensive intense sources of coherent light, suitable for holography. But at present only a laser will do, and unfortunately the cost of your laser is the largest part of the total cost in setting up a lab.

There are many kinds of lasers. Some will make holograms and some will not. Some have special features such as long coherence length or an output beam in some color other than red, but these are quite expensive. The only practical laser for most uses is the helium-neon gas laser. It should have output power of one milliwatt or slightly less, up to four or five milliwatts. All of the holographic methods described in this book can be carried out with any such laser. The determining factor in your choice is the cost, which tends to go hand-in-hand with the power. The least possible is about $300, and to go over $1000 is to be spendthrift. A $300 laser is quite adequate.

If you buy from an established dealer who understands the needs of holographers, you need not fear making a choice. Chapter 13 explains some technical points about lasers and offers further advice.

Optical Apparatus

The pieces of optical apparatus needed are few in number, simple in nature, and inexpensive. You need two small, short focal-length lenses for diverging laser beams, one (two would be better) front-surface mirror for changing the directions of beams, a beam splitter, and a filter for reducing the intensity of a beam, although the latter item can be dispensed with in most work.

Specifications for these items and their approximate costs are discussed in Chapter 13. However, because you may be unfamiliar with beam splitters, they will be explained here. A beam splitter is used to convert a single laser beam into two well-separated beams. It can be a simple device consisting merely of a piece of glass with parallel faces. Figure 8.1 shows how it works. Three incident rays are shown in a laser beam falling on one face. Some of the light will pass through and emerge as represented by ray A, parallel to the direction of incidence. Some will be reflected from the front surface to become a ray such as B. Some will go through the glass, reflect from the back surface, travel back through the glass, and emerge as in the case of ray C.

The rays that are wanted are A and one or the other of B and C. If the glass is thick enough (1/4 inch or more) B and C will be separated spatially so that the weaker one can be blocked off by a piece of cardboard. It is easy to make beam splitters from heavyweight glass, or they can be bought at very low prices. More elaborate splitters, some with partially silvered faces, and some made to exceptionally fine degrees of optical precision, are available but are unnecessary for our purposes.

These are the essential items. There is one more that is not mandatory, but which is highly desirable. This is a *spatial filter*. If a laser beam spread out

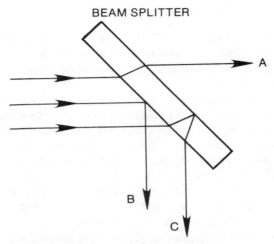

BEAM SPLITTER

A

B

C

Figure 8.1. *How a beam splitter works.*

by a lens is aimed at a wall, the illuminated region will almost unfailingly appear blotchy, instead of uniformly clean. The imperfections due to the lens can be reduced by making the lens impeccably clean or by using a better quality lens, but some of the blotchiness may be in the laser beam itself. A spatial filter placed at the exit port of the laser operates to remove the undesirable "noise" and to result in a clean, smooth, illuminated area on the wall. A cleaner laser beam means cleaner holograms.

Figure 8.2 shows how a spatial filter works. It consists of a short focal-length positive (converging) lens and an opaque plate with a tiny hole in it. The lens brings the laser beam to a focus and the hole is placed precisely at the focal point. The lens aims at the hole and only the wanted parallel rays get

Figure 8.2. *How a spatial filter works.*

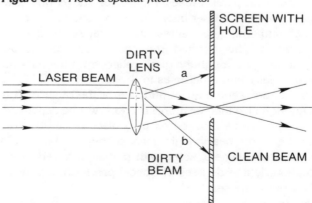

SCREEN WITH HOLE

DIRTY LENS

a

LASER BEAM

b

DIRTY BEAM

CLEAN BEAM

through the hole. Other rays, scattered from dirt or other defects in the optics, will be aimed by the lens so that they do not pass through the hole. Examples of such rays are shown as *a* and *b* in Figure 8.2.

A nice feature of a spatial filter is that it passes all of the parallel rays in the laser beam and cuts out only the noise. In practice, there is virtually no loss of intensity through use of the filter. Another nice feature is that the beam comes out of the filter already diverged, so that usually no other (and possibly dirty) lens need be used. On the other hand, spatial filters are expensive ($100 and up) and are not easy to make from scratch. See Chapter 13 for more information.

THE SIMPLEST ONE-BEAM TRANSMISSION HOLOGRAM

Figure 8.3 shows the simplest way to make a transmission hologram.[1] Simple though this "bounce-off-the-table" method is, an analysis of it turns up some important basic concepts in holography.

After the laser beam is diverged by the lens, some of the light goes directly onto the plate, and some reflects from objects on the table and then also goes on to the plate. There is therefore a reference beam (of which R is one ray) and an object beam (of which A, B, and C are rays), and these coherent beams produce the desired interference pattern in the plate. The shutter is inserted into the laser beam to permit the emulsion to be put in place, the shutter is

Figure 8.3. *Arrangement for a simple one-beam transmission hologram.*

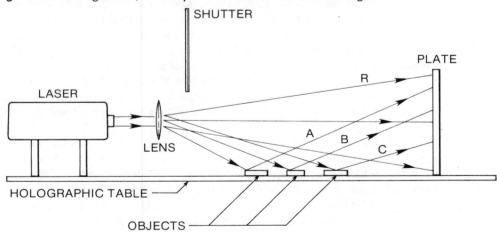

[1]This is an excellent starting point for beginning experimenters. A thorough discussion of it can be found in "The One-Beam Transmission Hologram" by Dr. T. H. Jeong, *The Physics Teacher*, February 1980, pp. 129–133.

lifted to expose the plate for a few seconds, and the shutter is reinserted so that the emulsion can be taken out for processing.

After the processed hologram is ready, it can be viewed by putting it back in position, removing the shutter and objects, and looking through the plate toward the light source. In the virtual image the objects will be seen realistically, with details discernible and with shadows and bright spots shifting as you move your head.

There are two principal limitations on the artistic quality of such a hologram. One is that the scene is strongly backlit, as though you were looking at objects at a shallow angle in the rays of the almost-set sun. The other is that the objects must be flat (Coins make excellent objects, being flat and reflective, and having interesting surface details.) Such a scene can have charm. For example, you can make a twilight desert setting with some sand and small figures. However, ordinarily larger and taller objects are preferred, with front and side lighting. This can be managed by a simple modification of the basic one-beam method, as will be discussed later.

Notice that considerable depth of scene is possible. If the central ray of the reference beam is not far from the level of the table top, the length of any ray in the object beam will not differ greatly from that for any ray in the reference beam.

INTERMODULATION NOISE AND BEAM RATIO

If you are listening to a record player or radio, you recognize disturbing extraneous sounds as "noise." Similarly, in viewing a hologram you may see blemishes in the image, and by analogy these can be referred to as noise. Some sources of noise are dirt or scratches on optical components, or water spots on the film. These can be eliminated by using good components that are kept scrupulously clean, and by careful processing of the emulsion. Another source that is certain to be present is the laser beam itself. Blotches that show up when the diverged beam is shone onto a screen will show up in holograms. This should not be very disturbing, but for best results a spatial filter must be used to clean up the beam.

Let's turn to a more fundamental source of noise. Consider Figure 8.3 again. We want the reference beam and the object beam to interfere. For example, we want each of the rays A, B, and C to interfere with the reference beam. But would not rays A and B interfere *with each other,* and similarly for A and C, and for B and C? Indeed they would, and the result would be unwanted interference fringes in the hologram. This *intermodulation noise* was discussed in Chapter 7.

The amount of intermodulation noise in a hologram is affected by the relative intensities of the reference and object beams, or *beam ratio,* and by the geometry of the configuration used in making the hologram.

Suppose that the intensities of the light associated with rays A and B are low relative to the intensity of the reference beam. Then the interference of A and B will produce fringes of low-maximum intensity. The usual silver-halide photographic emulsion is nonlinear in its response to light, and it does not begin to react strongly until the intensity of light on it rises sufficiently. Thereafter the darkening of the film will be in proportion to the light intensity. (Assume that the time of exposure of the emulsion is the same in all cases.) The consequence of the low amplitude interference effect and the nonlinearity of the film is that the amount of intermodulation noise recorded will be low. The interference of such rays as A and B individually with the reference beam, which has high intensity, will produce intensity variations on top of an intense beam.

Too great a ratio between the intensities of the reference and object beams is undesirable because the result would be a hologram with too little contrast between the maxima and the minima in the recorded surfaces. In fact a ratio of 1:1 would be desirable, if intermodulation noise were not a consideration. With some configurations for making holograms, the geometry works against intermodulation noise, and then a ratio of 1:1 may be desirable. Besides the influence of the geometry on the optimum beam ratio, the properties of the emulsion play a role, as indicated above.

In spite of these problems it is possible to specify numerically a beam ratio that is likely to give good results for a given kind of configuration, assuming use of the most common kinds of emulsion. For the one-beam transmission hologram method, the ratio is generally taken to be 4:1. When actually making a hologram, finding the best ratio is a matter of trial and error. This may sound like an unpleasant and time-consuming matter, but fortunately it can be ignored in practice, unless you find you are getting very poor results. Almost always the quality of the hologram does not depend critically on the ratio. Thus if 4:1 is recommended, 5:1 or 3:1 will work well, though 10:1 or 1:1 can be expected to be too far off. Furthermore, the ratio can be judged by eye well enough for most purposes.

CONSIDERATIONS ABOUT THE GEOMETRY

We pointed out earlier the importance of equalizing the path lengths for the reference and object beams. Now let's look at some other important features in the geometry of a holographic configuration.

In Figure 8.4 the solid curves indicate the nature of the surfaces that will be recorded, which are due to interference between the reference beam from the source S and light reflected from such points as O_1 and O_2 on the object. The dashed curves indicate the nature of the surfaces due to interference between beams from points O_1 and O_2. The latter surfaces give intermodulation noise during reconstruction of the image, but because they form large angles

S •

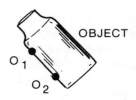
OBJECT

O_1

O_2

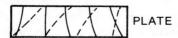
PLATE

Figure 8.4. *The interference fringes due to points O_1 and O_2 (dashed curves) are oriented differently from the fringes (solid curves) due to interference between the reference beam and the object beam.*

with the desired surfaces, they send their light off in directions far from the direction from which you view the image. It is best to avoid large objects close to the plate. For example, if point O_1 were farther to the left than is shown in Figure 8.4, the interference fringes that it and point O_2 produce in the plate would form smaller angles with the main fringes.

In Figure 8.5, consider angle *a* between ray R in the reference beam and ray B in the object beam from the scene, and also angle *b* between rays from points in the scene. It is desirable to have *a* large compared with *b*. In order to keep the noise out of the image, the angle between the reference and object

Figure 8.5. *The angles (a and b) associated with intermodulation noise.*

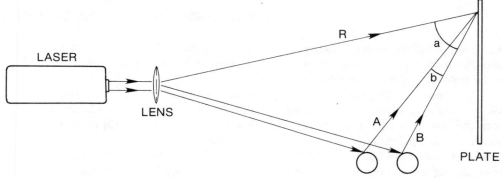

beams should be large, and the angular spread of the object beam should be small.

SPECKLE

Consider Figure 8.3 once more. Suppose that the apparatus is set up, but that the film is not in place. If you remove the shutter and look at the scene from the right you will see the objects as you will eventually see them in the virtual image. When you do this, you will see the coins or other objects and the table top showing a curious granular appearance, with a subtle scintillation or sense of variation in the granularity. Since this is what you can expect to see in the image given by the hologram, it will come as no surprise to see a similar *speckle*, as the effect is called, in the ultimate image.

If you look at the coins or other objects in ordinary incoherent light, you will not see any speckle. For that reason, you can regard the speckle in the image as a kind of noise. Actually the speckle is *meaningful* information about the surface structure of the objects.

Speckle is observable whenever a laser beam reflects from a surface to your eyes. A simple way ot observe it is to direct the beam to a wall. In looking at the illuminated region you see the scintillating granularity. It is best to diverge the beam with a lens to provide a sizable spot to examine, but this is not necessary.

This speckle is an interference effect. The beam of laser light is a coherent beam. When it reflects from some surface, variations in the texture of the surface will introduce into the reflected beam components with various phases. For example, a ray reflected from a small pit in the surface, and a ray reflected from a nearby raised region in the surface, will follow paths with different lengths to any point in the surrounding space. Where the rays are superimposed, interference will take place. The space can be thought of as filled with the speckle interference pattern. Your eye focuses the light onto your retina, and you see the interference pattern. If the reflecting surface is lightly textured, the details of the pattern will be fine-scaled, and you get the impression of a granularity. Finally, if you move your head, your eye will sweep over the interference pattern, and you will sense variations in the granularity—or what we call scintillation. Since you cannot avoid small movements of your eyes, the scintillation is sure to be seen.

An interesting corollary to this follows when you consider deliberately making large movements of your head. This gives you a sensation of a generalized movement across the illuminated region. A nearsighted person sees the motion from right to left as he moves his head from left to right. A farsighted person sees the motion from left to right under the same conditions. This is because the two kinds of eyes focus the light in different positions relative to their retinas.

Speckle is optical information about surface texture and could be considered to be desirable in a hologram. If you want a hologram that shows objects as they ordinarily appear (no speckle), then speckle is a blemish. It is the resolution of the film that determines how much speckle will be seen in a scene reconstructed with a hologram. Because speckle is due to surface details on a very fine scale, high resolution film is needed to record them. Low resolution film smears together the fine details in speckle, and the effect is not seen. You may agree that it is too bad that the effect of resolution is not the other way around, if sharp images without speckle are what you want.

Speckle has been studied much, and an entire branch of optics called "speckle interferometry" exists. International conferences have been held, and entire books written, dealing with this topic.

AN IMPROVED ONE-BEAM
TRANSMISSION HOLOGRAM METHOD

Figure 8.6 shows a one-beam method that provides for front lighting of the scene, allows use of taller objects, and permits some control over the path lengths, angles, and beam ratios. An actual setup of this kind is shown in Figure 1.4. The key feature is the front surface mirror. Because of the mirror, the reference beam and object beam travel different routes to the plate, and the two beams can be manipulated individually to some extent.

The recommended beam ratio is 4:1. After all else in the setup is as you

Figure 8.6. An improved one-beam transmission hologram method.

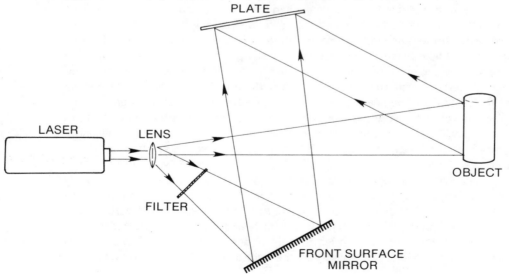

want it, it may be that this condition is not met. Most likely the object beam is too weak relative to the reference beam. In that event, an absorptive filter can be inserted in the reference beam.

A CYLINDRICAL TRANSMISSION HOLOGRAM METHOD

An especially striking kind of one-beam hologram is a *cylindrical* hologram. It gives a completely all-around, or 360 degree, view of the object, is easy to make, and tends to give sharp and brilliant images. Because it gives a view of the object from all perspectives, you could say that it is even more "holo-" ("complete") than flat holograms, which give only essentially front-on views.

A basic configuration is shown in Figure 8.7 (a). A transparent cylinder such as a plastic tube or glass jar, or even a piece of opaque tubing, holds the film. An object is placed as shown. An overhead laser shines its beam through a diverging lens and onto the film and onto the object. Light reflected from the object interferes with light falling directly on the film from the lens. Because this gives a top-lighting of the object, a reflective base for the cylinder may be used. A mirror surface with convex upward is best. A beam ratio of about 4:1 is desirable. The configuration gives you little control over this, for only the distance from lens to object is adjustable, but in practice this is sufficient to give very good holograms.

After the exposed film has been removed, processed, and dried, it can be reinserted in the cylinder if that is transparent, or simply formed into a cylindrical shape and held that way with paper clips or bits of tape. Then the diverged laser beam is shone on it as during making of the hologram, and the image is viewed from outside the cylinder.

With the arrangement shown in (a) the object can be changed easily, if it has a solid base to sit on, but the rather massive laser must be held at a considerable height. This requires very rigid supports if the laser is not to shake. A better arrangement in this respect is that shown in (b). Now the laser can rest at a low level on blocks, wood bricks, or the like. Unfortunately, there is a problem in holding the object to be holographed on the axis of the cylinder. Usually this is managed by cementing the object in place on the cylinder base.

A TWO-BEAM TRANSMISSION HOLOGRAM METHOD

Figure 8.8 shows a two-beam transmission hologram method. This avoids the main limitations of the one-beam method, and offers much more freedom in adjusting the parameters in the setup. This freedom is introduced by the beam splitter, which divides the original laser beam into separate reference and object beams. These can be manipulated individually.

Equalizing the path lengths, adjusting the angular relationships, and get-

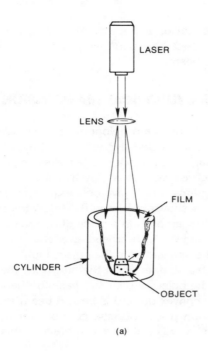

(a)

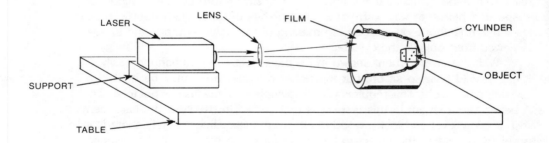

(b)

Figure 8.7. (a) Configuration for making a 360-degree cylindrical transmission hologram. (b) To avoid having to support the laser overhead, the setup can be horizontal on the holographic table top.

ting satisfying aesthetic effects are managed by changing the locations and orientations of the pieces of apparatus. The beam ratio can be assumed to be best at about 4:1. This may be achieved by adjusting the spread of the beams at the plate through movement of the mirrors. In addition, a filter can be used.

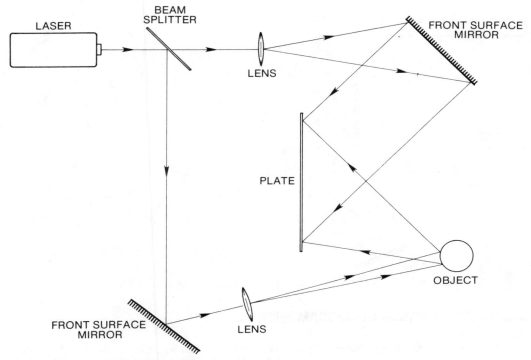

Figure 8.8. *A two-beam transmission hologram method.*

Since most often the object beam is much weaker than the reference beam, one filter in the reference beam should suffice.

SOME OTHER TWO-BEAM
TRANSMISSION HOLOGRAM METHODS

Still other configurations have been devised for making transmission holograms. Figure 8.9 shows one of these. Because each beam has its own lens there is still greater freedom in making adjustments in the setup, particularly with respect to the beam ratio, without using filters.

There is a considerable step up in complexity in Figure 8.9. The purpose is to open up the possibility of more interesting holograms from an aesthetic point of view. This is accomplished by illuminating the scene with two beams from different directions. This requires the use of two beam splitters, two mirrors, and three lenses. In practice the difficulty in getting a good hologram with this setup is not inconsiderably greater than with the simpler configurations.

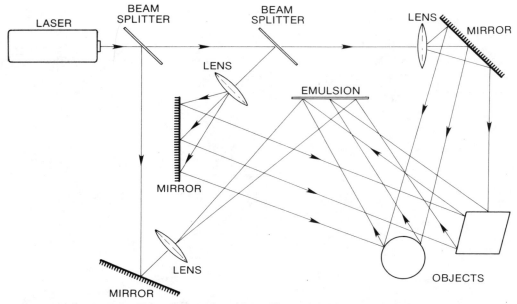

Figure 8.9. *How to have the objects illuminated from two different directions.*

A SIMPLE REFLECTION HOLOGRAM METHOD

Figure 8.10 (a) shows one of the simplest of all ways to make a hologram. Light in the diverged laser beam falls simultaneously directly on the emulsion from the left, and again on the emulsion from the right after reflection from the object. The reflection hologram that results is white-light viewable, and can give a sharp and bright image. The configuration makes the density of stacked surfaces high. Figure 8.10 (b) shows a modification of the method.

The beam ratio should be close to 1:1. The only control you have over this factor with this simple setup is through the choice of the object, which should be highly reflecting, preferably red or white in color, and through keeping the object close to the plate. If the object is substantial enough to permit it, the plate or film in its holder can simply be leaned back against it.

TWO-BEAM REFLECTION HOLOGRAM METHOD

Figure 8.11 shows a two-beam method for making a reflection hologram. The two beams from the beam splitter can be manipulated separately. By moving the lens in the reference beam in particular, the desired ratio 1:1 can be approximated. Highly reflective objects are best.

When the image is viewed, the object will be seen farther back from the hologram than with the previous method. In both cases the depth of scene is very limited.

110

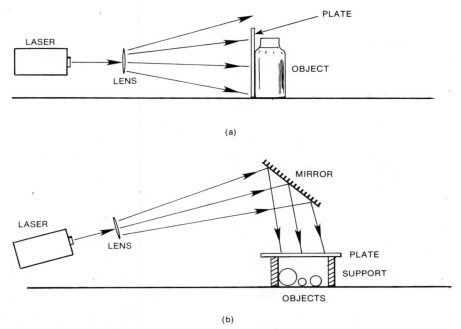

Figure 8.10. (a) The simplest configuration for making a reflection hologram. (b) A modified version.

Figure 8.11. A two-beam reflection hologram method.

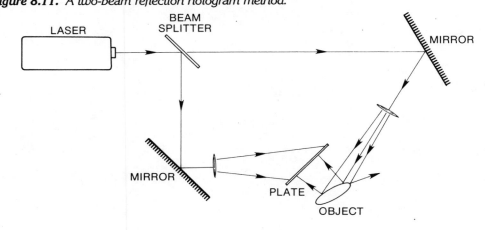

Diffraction and the zone plate model

The geometric model has been able to take us a long way in understanding holograms, but it has left some gaps. Holograms have properties that cannot be understood solely on the basis of the geometric model. You may wish to refer back to Chapter 7 for a commentary about this.

Besides the fundamental optical phenomena that the geometric model uses (interference of coherent beams, reflection of light from mirrorlike surfaces, and the Bragg effect) there are other optical effects that take place in holograms. These are grouped together under the heading of *diffraction phenomena*.

This chapter explains the nature of these phenomena and then uses the concepts to develop what is called the *zone plate model* of holograms.

The zone plate and geometric models, taken together, provide great power and flexibility in thinking about holograms. Some aspects of holography are most simply dealt with by using one or the other model, and some by using the two to complement each other.

WHAT IS MEANT BY DIFFRACTION?

Let's start by considering Figure 9.1, where (a) represents in a highly schematic way light falling on a thick transmission hologram. The light reflects from the surfaces (this being the heart of the geometric model), but also the incident

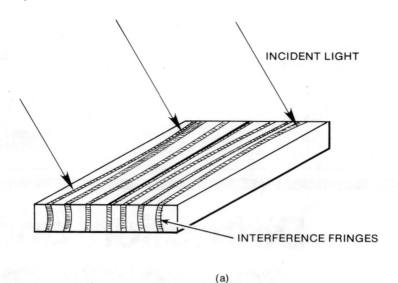

INCIDENT LIGHT

INTERFERENCE FRINGES

(a)

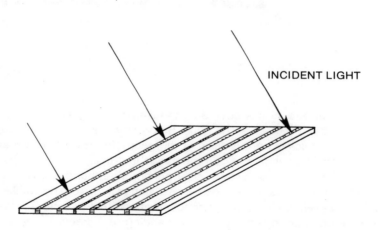

INCIDENT LIGHT

(b)

Figure 9.1. *Schematic representation of light falling on a thick transmission hologram (a) and on a very thin transmission hologram (b). In each case the light encounters slits and diffraction takes place.*

light encounter slits when it impinges on the hologram. It is this circumstance that gives rise to the phenomenon of diffraction. In part (b) of Figure 9.1 the hologram is so thin that the geometric model does not apply, yet diffraction is caused by the slits; indeed, diffraction must account for all the properties of such thin holograms. Thus you can see something of the importance of diffraction of light by slits in holography.

In Figure 9.2 a beam of parallel rays of monochromatic and coherent light is traveling to the right. It falls on an opaque screen in which there is an open slit. You might then assume that what happens is what is shown in the drawing: Rays that strike the screen are blocked, and those that reach the slit go on to the right without deviation. As a result you would expect a sharply defined beam of parallel rays at the right, bounded by darkness above and below. The edges would cast perfectly sharp shadows.

Part of what *actually* happens is shown in Figure 9.3. Rays that pass through the slit change their directions of travel, and *the beam spreads out into the region where darkness would have been expected.* Later we will explain how the physical cause of this spreading into the "shadow" region can be understood.

In this *single-slit diffraction* (as in Figure 9.3) the light that is bent around the edges of the slit is less intense than is the light that falls on the slit from the left. The more the intensity decreases, the greater the amount of the bending. Thus the intensity of the light associated with the diffracted ray *a* is less than that for ray *b,* and that is less than for the central undeviated light. Instead of

Figure 9.2. *In the absence of diffraction the edges of a slit would cast sharp shadows.*

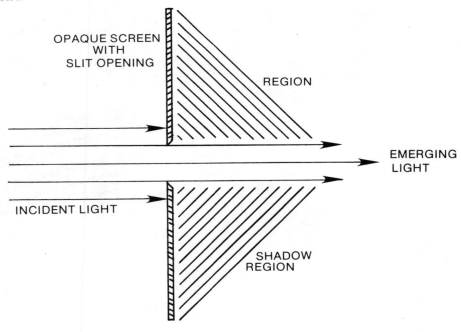

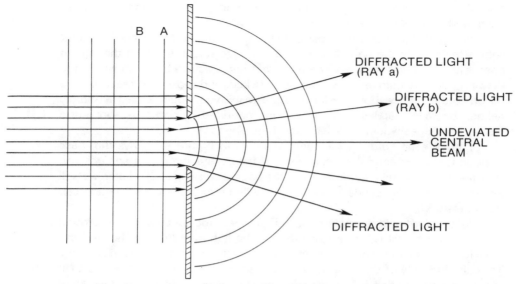

Figure 9.3. *Because of diffraction the wavefronts bend into what would otherwise be the shadow regions.*

the sharply defined beam of Figure 9.2, there is a spreadout beam that shades off in intensity to the sides.

There is one more aspect of this phenomenon that remains to be described. Figure 9.4 shows light from the slit that has been allowed to fall on a screen.

Figure 9.4. *General nature of the diffraction pattern with single slit. There are bands of alternating intensities, with the overall intensity dropping off on each side of the central region.*

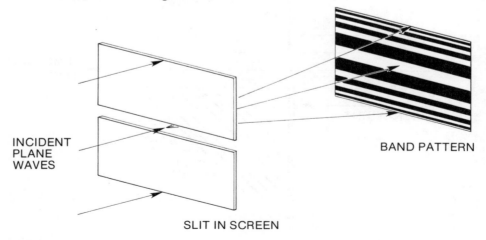

116

The illuminated region, besides being spread out, shows alternating bright and dark bands. The cause of these bands is mutual interference between diffracted rays.

Because of the kind of optical situation suggested in Figure 9.1, we began our discussion of diffraction effects with a description of single-slit diffraction. However, an even more basic effect occurs when light encounters a single edge, rather than a two-edged slit. A bending of the beam around the edge occurs, as shown in Figure 9.5. To distinguish this effect, it is called _edge diffraction._ The intensity of the deviated light decreases as the angle of deviation increases, so that ray _a_ is less intense than ray _b,_ for example. Also, alternating bright and dark interference fringes are produced.

Diffraction refers to the bending of light around any obstacle. Diffraction occurs not only when the obstacles have straight edges, as in the cases above, but for _any_ obstacle. For example, light entering the aperture of a telescope or microscope diffracts, and the image produced may be significantly affected by this. In fact, certain curved slits are much more important in holography than are such straight slits as those in Figure 9.1. The bulk of this chapter will deal with them.

If you have a laser and a lens with which to diverge the beam, you can easily observe the slit diffraction and edge diffraction effects. Shine the diverged beam onto a wall or screen and insert something with a sharp edge, such as a razor blade, into the beam before it hits the lens. You will see on the screen the spreading of the light into the "shadow" region, the drop-off in intensity with increasing angle of diffraction, and the presence of interference bands. By using two razor blades or the like you can easily produce slit diffraction and observe the similar effects. Anything with a hole in it, such as a washer with a small opening, or any curved opaque object, such as the head of a common pin, can be used with interesting results.

Figure 9.5. _Single-edge diffraction._

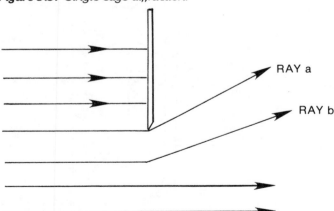

RAY a

RAY b

While diffraction can be very important in holography and in the science of optics generally, it is not a familiar optical effect in ordinary daily life. For example, we often look at light coming to us through a window (a kind of slit) or between the slats in a Venetian blind, but do not see a diffraction pattern. That's because slit diffraction is a pronounced effect only when the slit width is so narrow as to be comparable with the wavelength of the light. In the commonplace instances just mentioned, the slit width is enormous. On the other hand, the slit widths in a hologram (as in Figure 9.1) *are* comparable with a wavelength of visible light.

WHY DIFFRACTION TAKES PLACE

Let's return to Figure 9.3. Why do the light waves bend around the edges of the slit? The ideas of Christiaan Huygens can again provide a satisfying answer. The leading wavefront to the left of the slit is moving to the right. The parts

Figure 9.6. Each point in the slit becomes a source of Huygens' wavelets. The curve tangent to the wavelets is the new wavefront. Parts of the new wavefront travel into the "shadow" region.

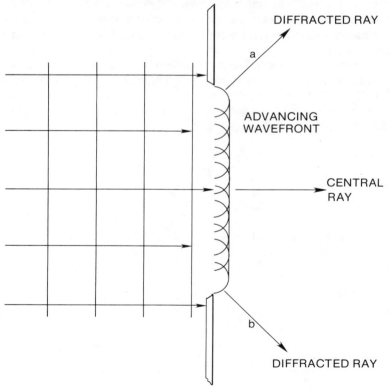

of the wavefront that hit the screen will be blocked, but the parts that arrive at the opening do something interesting. To help you visualize this, think of the waves as water waves in a ripple tank experiment. If there were a bobber in the opening, it would be agitated when the wavefront hits it. As it bobs up and down in response it will act as the source of circular wavefronts that spread out from it. This will happen similarly to every point along the wavefront in the slit. Thus Huygens' wavelets are created as is illustrated in Figure 9.6. According to Huygens' principle, the wavelets combine so that their "envelope" (the curve that is tangent to them) constitutes the new wavefront. This new wavefront is essentially straight over much of its extent, but it curves sharply at the upper and lower edges of the slit. To stress the bending of rays that accompanies this, two rays (labeled a and b) are drawn in. These move upward and downward into what would otherwise be the shadow region.

It is in this same manner that you can think of the cause of the slit diffraction of *light* waves, although such nonoptical details as the bobbers must be given up. Single edge diffraction, as in Figure 9.5, also can be understood by application of the Huygens' principle.

It is important to notice that the light rays falling on a slit need not be parallel. In Figure 9.7 there is a point source of light S that is near the screen, so that the wavefronts arriving at the slit are circular, rather than plane. Still, the wavefronts agitate the points in the slit openings much as before, though different in detail with respect to the shape of the new wavefronts that are formed.

Figure 9.7. *Diffraction also takes place at a slit when the waves arriving at the slit are circular, as is the case when the source S is near the slit.*

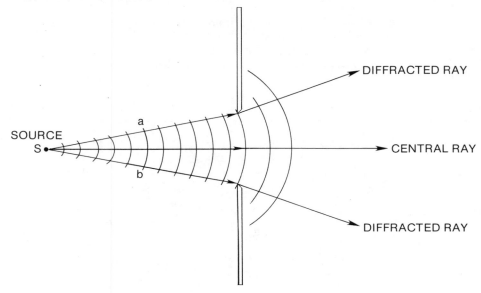

In this case there would be a well-defined shadow region if there were no diffraction, with the light beam spreading in accordance with extensions of rays a and b to the right. With diffraction, there is further spreading of the light into the regions above and below that shadow region.

Thanks to the idea of the Huygens' wavelets, diffraction (bending) of light past edges can be explained satisfactorily. The appearance of the interference bands in the beam of diffracted light (Figure 9.4) is not as easily accounted for, requiring somewhat detailed analysis of the phase and amplitude conditions in the diffracted light. You can find the analysis in any first-year physics textbook in the section on physical optics. This book will not go into this matter any further because in holograms the slits that produce diffraction are not single-slits with straight edges, nor even multiple-slit systems with straight edges, parallel, and uniformly spaced, as in the usual textbook treatment of multiple-slit diffraction.

ZONE PLATES

After these introductory discusions of some aspects of diffraction, let's turn to the kind of diffraction patterns that are produced when holograms are made. These patterns are simple in form and their effect on light passing through them is remarkable.

Figure 9.8 is an arrangement for making a transmission hologram of an object. Rays from a single point P on the object travel to the plate. This ray is a representative one in the beam from the object. Because point P is close to

Figure 9.8. *A plane reference beam and spherical object beam interfere at the plate, with an off-axis arrangement.*

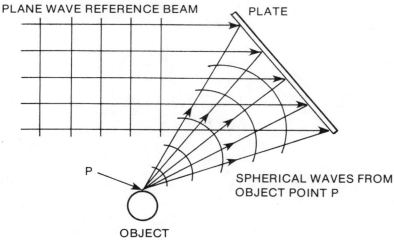

the plate, the wavefronts arriving at the plate are spherical in form. The reference beam, on the other hand, consists of plane wavefronts. Notice that this is very often the actual case in making transmission holograms, since usually the objects in the scene are much closer to the plate than is the effective source of the reference beam.

The important thing to notice in this is that the interference pattern that will be recorded in the plate is due to the interaction of plane waves with spherical waves. (It may be instructive to compare this with the situation in Figure 4.1, where the interference in the plate is between two sets of spherical waves.)

The configuration in Figure 9.8 is off-axis, and this is realistic. However, our discussion will be expedited if we turn instead to a strictly on-axis config- uration, as in Figure 9.9. Here the bundle of rays from point P form a right circular cone, and these are "cut" at the plate by the plane wavefronts of the reference beam. It is a matter of geometry that if you make a plane cut across a right circular cone you get a circle. Consider for example the rim of an ice cream cone. Then imagine you have a set of ice cream cones, nested into each other, and suppose that you were to make a cut across this stack of cones. The result would be a number of truncated cones with their rims forming a set of circles in the plane of the cut.[1]

The arrangement shown in Figure 9.9 results in a set of concentric circles similar to our ice cream cone example. It is not hard to see how this comes about. At point A the ray from P may be in phase with the reference beam, and there will be constructive interference at A. There also will be constructive

Figure 9.9. *An on-axis arrangement for interference between a plane reference beam and a spherical object beam.*

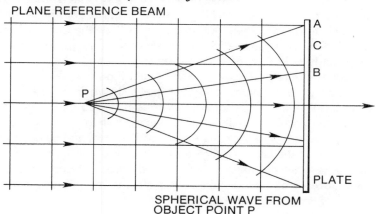

PLANE REFERENCE BEAM

SPHERICAL WAVE FROM
OBJECT POINT P

---

[1]The cut referred to is made at right angles to the common axis of the cones. If the cut were made at an angle to the axis somewhat different from 90 degrees the result would be a set of ellipses rather than circles. This can be the case with the off-axis arrangement of Figure 9.8.

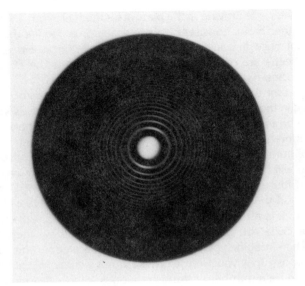

Figure 9.10. *A classical zone plate. In contrast with a "sinusoidal," or Gabor, zone plate, this one consists of concentric rings that are alternatively completely opaque and completely clear.*

interference at all points on a circle that passes through A and is centered on the axis. At some other point such as B there will also be constructive interference and a circle concentric with the circle through A, though with a smaller radius. At a point such as C (between A and B) there will be the out-of-phase condition and so the circle through C, centered on the axis, will be the locus of points of destructive interference. By extending this mode of reasoning, you can see how the overall result is a set of concentric circles in the plate.

The pattern of circles is called a *zone plate.* Figure 9.10 will give you a good idea of its general nature. The zone plate shown in Figure 9.10 is a particular kind called a *classical* or *Fresnel* zone plate. To make such a zone plate an artist draws concentric circles on a reasonably large scale with the radii of the circles proportional to the square roots of 1, 2, 3, 4, and so on, and then the artist blackens in the alternate regions. Finally, for use in a practical optical system, the zone plate the artist has made with large dimensions is photographically reduced. For example, the zone plate shown in Figure 9.10 would be about 1/2 centimeter in diameter.

A Fresnel zone plate would have a transmissivity profile like that shown in Figure 9.11 (a). Across the rings there are abrupt changes from blackness to transparency. The Fresnel zone plate is a diffracting device. If it is illuminated with a beam of light the circular open slits in the pattern will diffract the light as the light passes through them.

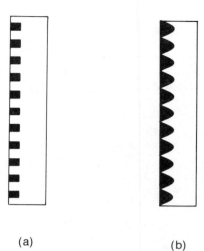

(a)   (b)

Figure 9.11. *The Fresnel and Gabor zone plate transmissivity profiles. (a) The classical Fresnel zone plate has abrupt changes from zero to full transmissivity. (b) The Gabor zone plate has a continuously variable "sinusoidal" profile.*

When the zone plate is made with the arrangement in Figure 9.9 with a photographic emulsion recording the interference pattern (as opposed to using an artist and subsequent reduction of scale of the drawing) what is recorded in the plate is a zone plate that differs from the Fresnel zone plate. The transmissivity profile varies in a smoothly continuous fashion from ring to ring, rather than abruptly at the edge of each ring. The nature of this profile is illustrated in Figure 9.11 (b). Mathematically speaking, the profile has the form of the square of a sine funtion, and the zone plate is often called a *sinusoidal* zone plate. Another common name is *Gabor zone plate*.

The importance of this distinction is that the classical and Gabor zone plates diffract light passing through them very differently. In holography the zone plates *are* recorded in photographic emulsions, and the prime question is how *Gabor* zone plates diffract light that is shone on them.

What happens is illustrated in Figure 9.12. The light beams spread out due to the action of the circular rings in the Gabor zone plate, and they fall on top of each other in the region to the right. Wherever this happens, the beams interfere. Surprisingly the result is that only *two* diffracted beams are produced. As is shown in Figure 9.12 one of the beams is convergent to the right, and gives a real, focused spot of light on the axis. There is also a divergent beam. If this is traced back as indicated by dotted lines in the drawing, it corresponds to a point on the axis to the left of the zone plate. This is a virtual source of the divergent beam from the zone plate.

The real focused spot and the virtual spot on the axis are respectively a real image and a virtual image of the point P in Figure 9.9.

Recognizing this gives one an "Aha!" feeling, since there is clearly a relation between this and the production of real and virtual images with holograms. However, you must take two more steps before the relationship is sufficiently complete. First of all, you must leave the on-line arrangement and return to the

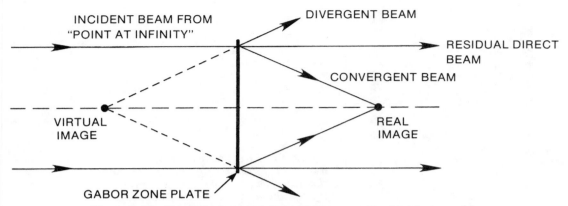

INCIDENT BEAM FROM "POINT AT INFINITY"

DIVERGENT BEAM

RESIDUAL DIRECT BEAM

CONVERGENT BEAM

VIRTUAL IMAGE

REAL IMAGE

GABOR ZONE PLATE

Figure 9.12. *The Gabor zone plate produces one virtual and one real image.*

off-axis arrangement of Figure 9.8. In such a case the cones of rays from the point P are intersected *obliquely* by the plane wavefronts of the reference beam. The result is an interference pattern resembling that of Figure 9.10 but with an elliptical form rather than a purely circular form. When this kind of zone plate is illuminated, a real and a virtual image of point P are reconstructed, but now the images are off the axis of the system. The virtual image lies at the original location of P, and the real image at a symmetrically located off-axis point to the right of the plate. Figure 9.13 will help to make this clear.

Figure 9.13. *The images can be well-separated by using an off-axis arrangement.*

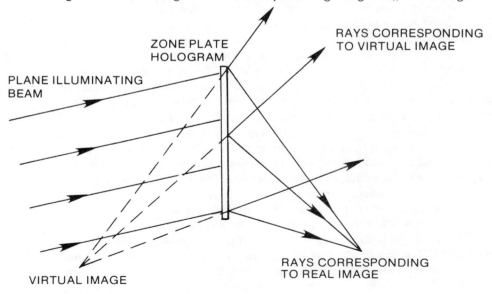

ZONE PLATE HOLOGRAM

RAYS CORRESPONDING TO VIRTUAL IMAGE

PLANE ILLUMINATING BEAM

RAYS CORRESPONDING TO REAL IMAGE

VIRTUAL IMAGE

124

Finally, you must make a mental leap to adapt this picture to the case of the whole object in Figure 9.8. Each point such as P on the object will produce its own zone plate in the emulsion, and the whole hologram that results will be a complicated collection of zone plates. There will be one from each point on the object and each will be different in detail from the others. When the hologram is illuminated with a reconstructing beam, real and virtual images will be produced of each of the points on the object. In each image the reconstructed points will have proper spatial relationships with each other and so be fully three-dimensional. In addition points of various degrees of brightness in the object will yield zone plates with rings of various degrees of opacity alternating with various degrees of transparency. Therefore the reconstructed points will be seen with appropriate degrees of brightness. In short, realistic real and virtual images of the object will be reconstructed.

INTERLUDE ON FRESNEL ZONE PLATES

This section is addressed to those who may want to dig into physics books to learn more about zone plates than we have had reason to say here. Such books discuss classical (or Fresnel) zone plates almost exclusively, so the connection with the Gabor zone plates that appear in holograms made with photographic emulsions is not to be made without modification.

The usual way in which Fresnel zone plates are made was explained in the preceding section. Such zone plates have been known for many decades, and their properties have been studied mathematically and experimentally during that time. These zone plates produce not just one real and one virtual image, but numerous images of each kind. What happens is illustrated in Figure 9.14.

Figure 9.14. *The classical zone plate gives numerous virtual images and numerous real images.*

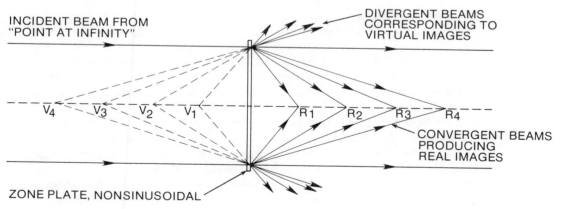

For simplicity the figure shows only four virtual images (V_1, V_2, V_3, and V_4) and four real images (R_1, R_2, R_3, and R_4), but in principle an infinite number of images of each kind are produced.

Such a zone plate would not be useful in holography because of this great multiplicity of images. In viewing the scene you would see a confused jumble of images. But with sinusoidal zone plates there are only one real and one virtual image of each point in the scene. It is not only the concept of alternating black and clear rings in zone plates (as in Fresnel zone plates) that is important in holography, but also that the photographic process gives the rings a "sinusoidal" transmissivity profile.

GABOR ZONE PLATES AS LENSES

Some glass lenses are thicker in their centers than near their outer edges, and converge light to form focused images. Such a "positive" lens is the kind that can be used to condense sunlight to form an extremely hot spot. Other lenses are thinner in their central regions than near their outer edges, and these are diverging, or "negative" lenses. A person who is nearsighted would use spectacles with lenses of this kind.

A zone plate hologram contains individual Gabor zone plates and each produces both a diverging beam and a converging beam. Therefore you can think of the zone plates as simultaneously positive and negative lenses. To adopt this view is simpler than to think of the more complicated ideas of diffraction by complex rings and interference in the diffracted beams, and this can be a very useful picture to adopt. However, you should not forget that analogies must not be pushed too far when drawing conclusions from them.

COLOR EFFECTS WITH GABOR ZONE PLATES

In our discussions of edge diffraction and of zone plates, until now we have either explicitly or implicitly assumed that monochromatic (possibly laser) light has been used. For example, if a zone plate hologram is made and then illuminated for viewing using the usual red laser light, our discussions apply. But suppose that the hologram were made with laser light and then illuminated with light of some other color, or with white light. What happens is the subject we take up now.

In Figure 9.15 (a) the light falling on the zone plate has comparatively short wavelength—blue light, for example. In (b) the wavelength is relatively long—red light, for example. In each case a real image and a virtual image (not shown in the drawings) are produced, but at different distances from the zone plate. Specifically, the blue light is diffracted (bent) less than is the red light, and so the blue image lies farther from the plate than does the red image.

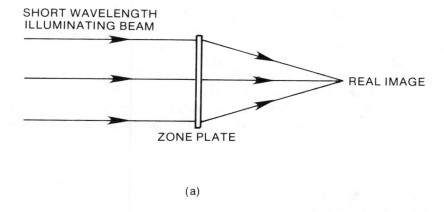

(a)

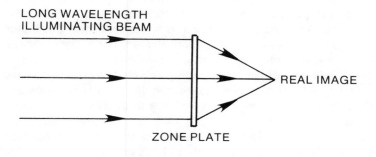

(b)

Figure 9.15. *The location of the image depends on the wavelength of the light. Short wavelength light (a) and long wavelength light (b) put the images farther from and nearer to the plate, respectively.*

If a zone plate hologram made with red laser light is illuminated with a reconstructing beam that contains both red and blue components, *four* images would be seen. Two of these would be real images, one red and one blue, and two would be virtual images, one red and one blue. These images would be displaced spatially from each other along the axis. If the hologram were off-axis instead of in-line as in the drawing, the images would be displaced sidewise as well as lengthwise.

Now suppose that the illuminating light is white light, containing components of all spectral colors. With an off-axis hologram of an extended scene, the picture would be viewed as color-smeared sidewise. Also, the variously colored pictures would be offset laterally, since they would be formed at different distances from the plate. They would also be of slightly different sizes, because magnification by a plus-minus zone plate "lens" is different for different wavelengths, as is true of ordinary glass lenses, where the effect is called "chromatic aberation."

The effects are easily seen if you have an ordinary transmission hologram at hand and look at it with sunlight or with a bright single roomlight. Under such conditions you can verify that the variously colored versions of the scene are not of the same apparent size, nor evidently located at the same distances from the plate.

SUMMARY AND COMMENTARY ABOUT THE ZONE PLATE MODEL

A zone plate hologram is thought of as a very complicated collection of Gabor zone plates, one for each point in each object in the scene that was holographed. The zone plates vary with respect to their locations and orientations, the degree to which they approach or deviate from circularity, the radii of the rings in them, and the degree of blackening of their dark rings. When the hologram is suitably illuminated with a readout beam, the zone plates produce both real and virtual images of the corresponding points in the objects, in their proper positions and with proper brightnesses, to give in the ensemble a three-dimensional representation of the scene.

The action of each zone plate is attributable to the fact that it contains curvilinear slits that diffract the readout beam of light, and to the interference of the spreadout diffracted beams from the slits. A simpler though less sophisticated view is that each zone plate is simultaneously a converging and a diverging lens. The vast collection of such lenses in the hologram, with their various locations, orientations, and focal lengths collaborate to produce the virtual and real images of the scene.

A zone plate diffracts light differently for various wavelengths. Thus, some of the color effects seen when some holograms are illuminated with light that contains components with various wavelengths can be understood. Readout with a beam of white light is a special case, resulting in color-smeared images.

For some holograms the geometric model is not applicable. Yet diffraction by the fringes recorded in the hologram takes place and the zone plate diffraction model succeeds. The case of the very thin transmission hologram is an instance of this.

There are still other useful results that can be drawn from the zone plate model. For example, by regarding the zone plates as lenses, the study of the vertical (crosswise) and horizontal (lengthwise) magnifications in the images is expedited. As another example, if you start with a zone plate such as that shown in Figure 9.10 and make each white ring black and each black ring white, the complementary zone plate that results acts in the same way as did the original. A practical consequence of this is that a photographic copy of a hologram can be made and will work, although in the copying process there will be losses of details and some introduction of noise.

Sometimes either the geometric model or the zone plate model may be preferable in explaining properties of holograms. There are instances in which the zone plate model is better and even completely necessary. An instance in which the geometric model is better is supplied by reflection holograms. Surely a reflection hologram consists of zone plates, but to think of it that way does not offer anything like the simplicity of the view that it is made up of stacked-up semireflecting surfaces that give rise to the Bragg effect.

"DIFFRACTION GRATING" HOLOGRAMS

Figure 9.8 shows how light reflected from a single point on an object near the plate interferes with the plane reference beam to produce a Gabor zone plate pattern. With this off-axis arrangement the rings in the zone plate are not circular, and the center of the pattern may lie off the plate. Then the pattern in the plate consists of curved arcs alternating sinusoidally between darkness and lightness. Even though this interference pattern is a segment of a full zone plate, it acts essentially as would a full zone plate. When it is illuminated with a readout beam it forms a real image and a virtual image of the point.

Now think of point P as being moved farther and farther away from the plate. As this happens, the curved arcs in the interference patterns become more and more straight. At the same time they become more narrow and more closely spaced. To see why these effects happen, look at Figure 9.10 and compare regions of the pattern near the center and near the outer edge.

In the ultimate limit of the regression of the point P from the plate, both the object beam and the reference beam become plane wavefront beams. The situation is illustrated in Figure 9.16. The interference pattern that will be recorded in the plate consists of straight, parallel, and uniformly spaced straight lines. Very appropriately, this is called a *grating,* or (since the pattern will be a diffracting device) a *diffraction grating.* As should be expected, the transmissivity profile in the developed emulsion will be "sinusoidal," as in Figure 9.16 (b). A grating with this feature is called a *Gabor grating.*

What is the relation between these observations and actual holograms? If the object point P is close to the plate when a hologram is made, the finished hologram is distinctly a zone plate hologram. With the object point farther away, the hologram begins to resemble a "straight-ruled" diffraction grating. If we think of the point as being infinitely far away, a true grating is produced. We have become rather abstract when we speak of a hologram of an infinitely faraway point. If a hologram gives an image of a scene, it is a zone plate hologram. Nevertheless, there is a resemblance between the properties of a

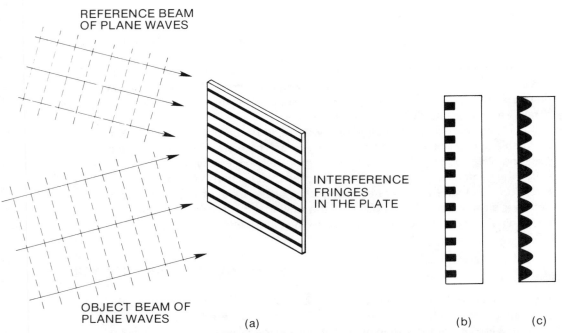

REFERENCE BEAM
OF PLANE WAVES

OBJECT BEAM OF
PLANE WAVES

INTERFERENCE
FRINGES
IN THE PLATE

(a)

(b)

(c)

Figure 9.16. (a) *The interference of two plane-wave beams produces straight parallel fringes, and the pattern in the plate is a diffraction grating.* (b) *The grating actually recorded in photographic emulsion has a sinusoidal transmissivity profile, as does a Gabor zone plate.* (c) *The classical diffraction grating has a profile like that of a Fresnel zone plate.*

zone plate hologram and a Gabor grating, and more people are probably familiar with gratings than with Gabor zone plates.

A BRIEF INTRODUCTION TO GRATINGS

Figure 9.16 (c) shows the transmissivity profile of a classical grating—the kind discussed in most physics textbooks. Abrupt changes from transparency to blackness take place across the "rulings" of the gratings. (The stripes are called rulings because for a century or so gratings were made literally by ruling a plate of glass with a diamond-tipped tool.) Such a classical grating is by no means an obscure optical curiosity. In fact it has played such an important role in the development of modern science, and continues to play such an important role in today's science and technology, that its fundamental importance can not be exaggerated. Even so, we will restrict our discussion of gratings to a summary of the most basic ideas, and the relatively slender connection between true straight-ruled gratings and holography. On the other hand, it is not difficult to make gratings with the arrangement shown in Figure 9.16. If you are interested and have the basic laboratory facilities described in Chapter 13, you can make excellent and very interesting gratings. The Gabor gratings that result could be

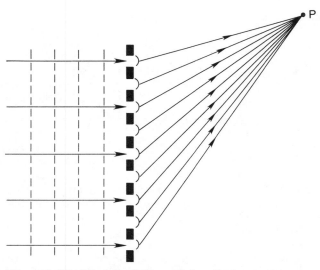

Figure 9.17. The light falling on the slits in a grating is diffracted by each slit, as suggested by Huygens' wavelets in the drawing. At a point such as P, rays from each of the slits will be superimposed and will interfere.

studied in comparison with classical gratings, and used for Science Fair projects and undergraduate college research projects.

In Figure 9.17 plane waves of monochromatic light fall on a grating. Huygens' wavelets spread out from each slit in the grating, and the multiple beams of spreading light from the slits superimpose themselves everywhere, such as at point P. The superimposed beams have different phases at P because of the different path lengths they travel from slit to P. The interference at P may be constructive or destructive, or something between these two extremes.

Although it is not immediately obvious, the result is that the grating accepts from the left in Figure 9.18 a single beam of one-color light, and produces off to the right more than one beam of the same color. Figure 9.18 assumes a classical grating, and as is shown, there will be several beams above and below the axis. These will appear in pairs, as for example one at the angle *a* up from the axis and a corresponding one down at the same angle *a*. The number of beams and their angular deviations depend on the slit widths, the slit separations, and the wavelength of the light used. In the case illustrated there are six diffracted beams and a residual direct beam along the axis.

While this production of multiple beams contrasts with the production of only a single real image and a single virtual image by a hologram, remember that the grating has been assumed to be a classical grating. Figure 9.19 shows what a *sinusoidal* grating does, such as would be made with photographic recording as in Figure 9.16 (a). The result is also not immediately obvious, but

131

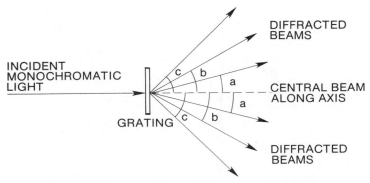

Figure 9.18. The classical grating accepts a beam of monochromatic light and produces a set of distinct beams (called "orders"). These are pairwise symmetrically located above and below the grating axis.

it is extraordinarily simple. This kind of grating produces only two diffracted beams, symmetrically directed up from and down from the axis, where a residual direct beam flows. On learning this you are again tempted to say "Aha!" and associate the two beams with the virtual and real images. But the diffracted beams are plane-wave beams, and we are dealing here with objects "at infinity." The conclusion oversimplifies what is really a more subtle situation.

In Figure 9.20 (a) the light falling on the sinusoidal grating contains a red component and a blue component. Since the diffraction of light by all gratings is different in detail for different wavelengths, you could expect that the two colored components would be diffracted differently. In fact, four beams are diffracted, with a red one above and a red one below the axis, and a blue one above and a blue one below the axis. If the incident light is white light, the effect is that a spectral smear of the light above and another below the axis is produced. A classical grating behaves similarly, but again the distinctive feature of the Gabor grating is that only two of the continuous spectra are produced, while the classical grating gives more than two, and even a substantial number. The nature of the two spectra from a Gabor grating is illustrated in Figure 9.20 (b).

Figure 9.19. The sinusoidal diffraction grating produces only two diffracted beams.

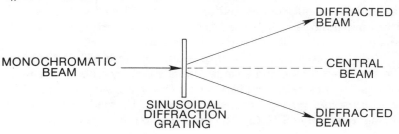

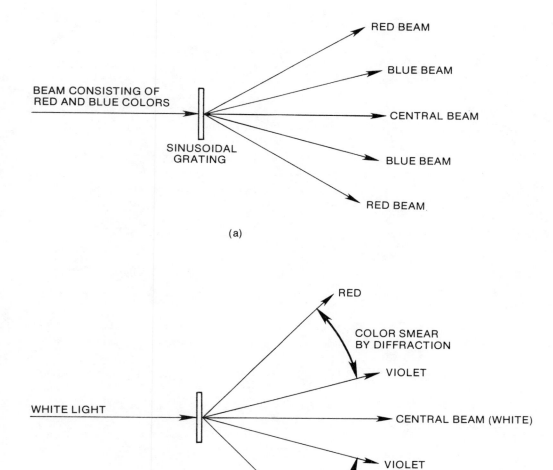

Figure 9.20. *(a) A mixed beam of two colors results in four diffracted beams, two in each color. (b) A beam of white light results in two continuous spectra.*

Rainbow and other "copy" holograms

Let us summarize what the basic scheme in holography is. Light from an object carries optical information about the object. The information is in the phase and amplitude of the light. A record of both of these aspects of the light can be made by adding a reference beam, so that an interference pattern results. At any later time an image of the object then can be reconstructed by shining a suitable readout beam onto the recording.

In spite of the simplicity of this scheme, there can be numerous variations in how it is carried out in practice. For example, certain changes in the relative positions and orientations of the reference beam source, the object beam source, and the plate correspond to changes from transmission to reflection holograms. Much ingenuity has gone into devising still other ways of making holograms. Some of the special types that have been developed are important state-of-the-art holograms today, and it is these that are discussed in this chapter. Still other special types have been proposed but await further development before they become practical. Among these are some of enormous potential importance, such as holographic movies and television, to be discussed in Chapters 11 and 12.

The holograms we will describe and explain now are referred to as *hybrid* types. They are holographic copies of other images. In some cases they are holograms of holograms, and in others, holographic images of two-dimensional pictures. This accounts for the use of the word "copy" in the title for this chapter.

135

PROJECTION HOLOGRAMS

Perhaps the simplest way to convey what a *projection hologram* does is to describe a typical one. Suppose the hologram shows a model of a dinosaur. With light falling on the plate from behind, you see the forepart of the body with the long neck and small head projecting forward from the plane of the hologram, and the rest of the dinosaur extending out from behind the plate. The effect can be very striking. If you move your head about, the dinosaur will be seen from different perspectives with great realism. A projection hologram is also known as a *standout hologram,* or as an *image-plane hologram,* the latter name referring to how the plate intersects the image. It is also possible to make a projection hologram with *all* of the image seen as floating in front of the plate. Not only are such holograms dramatic visually, but they are interesting from a theoretical point of view. Furthermore, the ideas they incorporate are used in making still other kinds of hybrid holograms, as you will see later.

Making an image-plane hologram is illustrated in Figure 10.1 (a). The plate that will be the hologram is being exposed to a reference beam and to an object beam, as is always necessary. The trick is that the object beam does not come from a real object, but rather it is light that comes from a "master" hologram and forms a real *image.* While the copy plate could not be placed to intersect a real solid object being holographed, it can be placed to intersect the real image of the object.

Thus the real image from the master hologram is being used as a stand-in for a real object. This is a useful brief description, but there is an important way in which this is not true. In order to understand the difference, suppose you are to make a projection hologram of an actual model by slicing the model and placing the parts up against the plate, one part on each side. This is not an equivalent method, because in Figure 10.1 the scene being holographed is entirely pseudoscopic, but the halves in the actual model are both orthoscopic. (There is also a problem about illuminating the split model suitably.)

When the copy hologram has been processed, it is illuminated in the usual way for transmission holograms, and as always there will be a virtual image and a real image. With an ordinary hologram the images are well separated and they are not seen simultaneously. With the projection hologram they *are* both seen, since they are very close together and in fact join at the plate. The observer sees the virtual image behind the plate and the real image in front of the plate.

The real image is a pseudoscopic version of the real image from the master plate, which is also pseudoscopic. Thanks to this double inversion, the real image that is seen is orthoscopic. The virtual image part of the scene is pseudoscopic, but one is not ordinarily aware of this, perhaps because of the overpowering clue given the observer by the connection of the two images at the plate. (Remember that perceiving pseudoscopy is a matter of *interpretation* by the observer.)

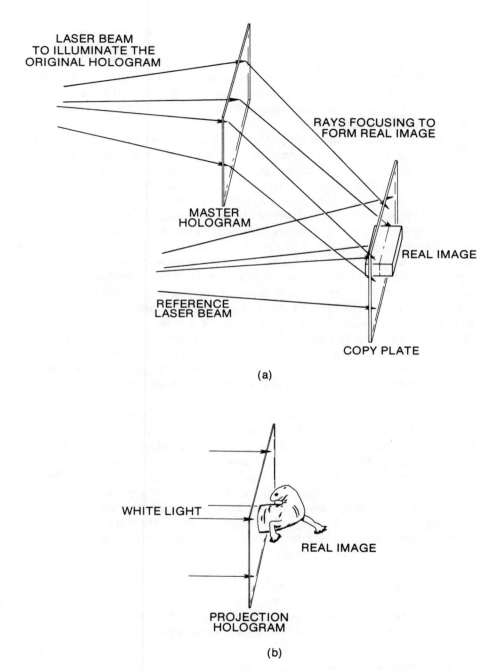

LASER BEAM
TO ILLUMINATE THE
ORIGINAL HOLOGRAM

RAYS FOCUSING TO
FORM REAL IMAGE

MASTER
HOLOGRAM

REAL IMAGE

REFERENCE
LASER BEAM

COPY PLATE

(a)

WHITE LIGHT

REAL IMAGE

PROJECTION
HOLOGRAM

(b)

Figure 10.1. *(a) Configuration for making a projection hologram. (b) To illustrate the text discussion of production of a black-and-white image.*

Projection holograms offer another advantage besides their unusual floating images. Although they are transmission holograms, they are viewable with white light without the usual smearing of the images. In fact, most of the image appears as *white,* although there is usually some color fringing in the outer parts. When the object portrayed in the image is white, the whiteness of the image contributes strongly to the realism. White plastic models, white chessmen, or white plaster statuettes make excellent subjects.

As an example of this effect, Figure 10.1 (b) shows a beam of white light falling on the projection hologram. As you know, a transmission hologram diverges the various colors present in white light and normally gives a color-smeared image. But when the image of the foremost part of the dinosaur is formed so close to the plate, the divergence of the various colors is slight. A similar argument holds for the part of the scene behind the plate.

Another aspect of projection holograms that distinguishes them from other transmission holograms is that the "object" (that is, the image of the object) lies so close to the emulsion that information about each point is not distributed over the whole emulsion. Instead, the interference fringes that pertain to any one point lie in the emulsion in the region that is very near the point on the object. For that reason the light used to view the image is acted on by the regions of the holograms much as it would be if it were shone directly onto the object itself. This fact also helps you to understand the white image seen with a projection hologram when white light is used. This correspondence between points on objects and small regions in a hologram of the objects occurs also with a reflection hologram made with the objects very close to the plate.

Until now we have been discussing *transmission* projection holograms. However, it is possible to make a *reflection* hologram copy of a real image from a master transmission hologram. Like other reflection holograms, such a projection hologram is white-light viewable. It also can give a standout image that is brilliant, white, and with considerable depth of scene.

Another projection hologram is called a *focused-image hologram.* This is not made with the real image produced by a master hologram, but instead by using a lens to make an image of a real object, with the holographic emulsion placed in this lens-produced image. Unfortunately this results in a pseudoscopic image when the hologram is viewed.

HOLOGRAPHIC LENSES AND REAL LENSES

Let's turn now to an interesting special application of the projection hologram idea. It involves real lenses and holographic images of lenses, and analysis of it is very instructive. This was originated in an article by Mac Rugheimer and Larry D. Kirkpatrick.[1]

[1]Mac Rugheimer and Larry D. Kirkpatrick, "Demonstration Holograph for Comparing an Image Lens and a Real Lens," *American Journal of Physics,* November 1977, pages 1027–1029.

First consider Figure 10.2. In (a) an observer is looking at the virtual image of a scene that consists of a sphere and cube. He or she has placed a real magnifying lens on the far side of the plate in an attempt to magnify the images of the objects. This will not work, because the lens is located where there are no real light rays from the objects for it to act on. The real rays that the observer uses to see the objects are the diverging rays that come to the viewer on his or her own side of the plate. The lens intercepts some of the laser beam that is being used to illuminate the plate, and so may affect what is seen by altering the nature of the readout beam at the plate, but it does not magnify the image.

In (b) the lens has been moved to the observer's side. Now it does intercept

Figure 10.2. _(a) The observer tries to use an actual lens placed on the far side of the plate to magnify the virtual objects. (b) Here he or she places the actual lens on his or her side of the plate. (c) There is in the holographed scene the objects and a lens. The virtual image of the lens acts on the virtual images of the objects as would an actual lens on actual objects. (d) Now there is both a virtual image of a lens in the scene, and an actual lens as well._

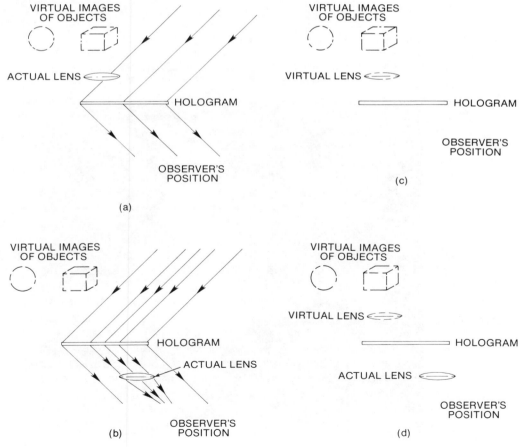

real light rays that pertain to the image, and it can affect their focusing. (Indeed, the observer in looking directly at the virtual image is using the real lens in his or her eye for the same purpose.) Notice that the phenomena in (a) and (b) can easily be tried if an ordinary transmission hologram and a simple magnifying lens are available.

In (c) the situation is distinctly different. When the hologram was made there was present in the scene not only the sphere and the cube but also a lens. Real rays in the object beam passed through the lens, and the transmitted rays that reached the plate were affected by the lens action. This was recorded in the holographic plate, including the fact that the amount of the effect depended on spatial relationships. When the observer views the image, he or she can move his or her head and look through the image of the lens at parts of the scene, or look around the lens and see the objects directly. When the observer looks through the lens, he or she sees magnified images. Figure 10.3 shows the effect for a lens and a stopwatch.

Figure 10.3. *The virtual image from a transmission hologram, showing a magnifying lens acting on a stopwatch. The image of the lens acts on the image of the watch as would an actual lens on an actual watch. If this scene were viewed "live," and from various directions, the parts of the watch that are magnified would change correspondingly. (Photo by Dr. C. E. Hamilton, produced by Mr. Rainbows.)*

In (d) of Figure 10.2 the observer is looking at the virtual image of the scene, and this includes a lens, as in (c). But there is also an actual lens, which the observer places on his or her side of the plate. Both lenses—the holographic lens and the actual lens—are capable of magnifying parts of the scene, and you might well conclude that if the observer lines the lenses up he or she will perceive joint action by the lenses, as if he or she were using a pair of actual lenses.

But the plot thickens. The lenses cannot cooperate as would two actual lenses because they act on different sets of light rays. The holographic lens acted on certain rays from the objects when the hologram was made, but in viewing the image the actual lens acts on different rays.

How can you use a holographic lens and a real lens so that they can act together as though they were both real? The problem can be solved by using a projection hologram. First a master hologram is made of the scene, which includes a lens. Then the master hologram is used to create a real image. Next, the real image is used to make a copy hologram. The arrangement is shown in Figure 10.4 (a).

When the copy hologram is illuminated for reconstruction as in (b), the observer can look through the image lens at the other objects and see them magnified. If an actual lens is also used, it acts on real light rays and can also produce a focusing action. The distinction between the holographic lens and the actual lens disappears as far as visual effects are concerned. You can look around both lenses and see the other objects in the scene directly, you can use one or the other lens singly, or you can put the lenses side by side and use them as in a binocular. In particular, you can *superimpose* them to form a compound lens—something that cannot be done with two actual lenses.

To quote Rugheimer and Kirkpatrick: "Notice that the image lens is so real in appearance that it fooled you until you tried to touch it. That image is so real that it fools the real lens into believing there is another real lens present. This demonstrates the faithfulness of the holographic recording process. One cannot help but be impressed."[2]

RAINBOW TRANSMISSION HOLOGRAMS

If a transmission hologram is viewed with white light, the image is seen as a smear, with spectral colors spread out across the smear somewhat as in a rainbow. While the name *rainbow hologram* might well be used for such a hologram, the name refers to a very special kind of transmission hologram that was developed by Stephen Benton in 1969. We will first describe the distinctive properties of a rainbow hologram, and then discuss how it is made and why it works as it does.

---

[2]Rugheimer and Kirkpatrick, "Demonstration Holograph for Comparing an Image Lens and a Real Lens," pages 1027–1029.

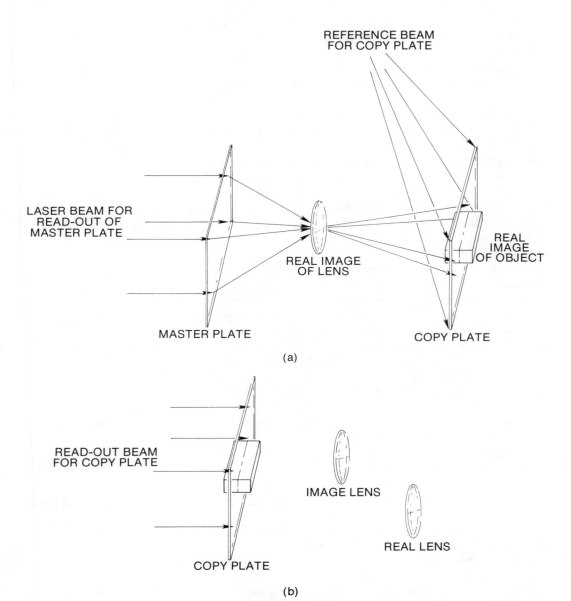

REFERENCE BEAM
FOR COPY PLATE

LASER BEAM FOR
READ-OUT OF
MASTER PLATE

REAL IMAGE
OF LENS

REAL
IMAGE
OF OBJECT

MASTER PLATE

COPY PLATE

(a)

READ-OUT BEAM
FOR COPY PLATE

IMAGE LENS

REAL LENS

COPY PLATE

(b)

Figure 10.4. (a) As when a projection hologram is being made, a copy hologram is made, using the real images of the lens and objects that are produced by the master hologram. (b) When the image is viewed using the copy hologram, the object can be magnified with the image lens, or with the actual lens, or with both simultaneously.

1. Suppose in Figure 10.5 that the illuminating beam is plane wave laser light of the same color as was used in making the rainbow hologram. If you, the viewer, put your eye at position a, you will see the image. It will of course be in the color of the laser light. It will be three-dimensional, yet restricted. If you move your eye from side to side, you will perceive full parallax in the scene, but if you move your eye up and down, you find that at positions such as b and c you cannot see the image at all. When your eye is near a, you are limited in changing your viewpoint vertically.

It is easy to simulate some of these effects by forming a horizontal slit with two fingers of one hand and looking at some object such as a book through this slit. When you see the book and move your head sidewise, you get full horizontal parallax in what you see. If you move your eyes too far up or down, you cannot see the book at all. The analogy is not perfect because when you lose sight of the book you will see other objects such as parts of the table, whereas with a rainbow hologram nothing is seen when the eye is too far up or down. Still, this little experiment shows nicely what is meant by "full horizontal parallax" and "loss of vertical parallax."

Looking at the scene with an ordinary transmission hologram is often (and correctly) described like looking at the scene through a window. Looking at the same scene with a rainbow hologram is like looking through a horizontal slit window.

2. Suppose the rainbow hologram were made with red laser light, but that it is being viewed with monochromatic light of some other color. The image will

Figure 10.5. Viewing a rainbow hologram. At and near the position a the image is seen. Farther out, as at b and c, it is not seen. At or near position a there is horizontal parallax in the image.

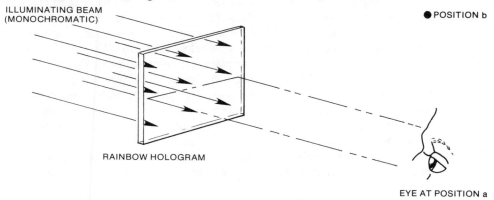

ILLUMINATING BEAM
(MONOCHROMATIC)

● POSITION b

RAINBOW HOLOGRAM

EYE AT POSITION a

● POSITION c

be visible in the color of the readout light, but you will have to shift your eye from position "a" upward or downward by a suitable amount. If the hologram is thought of as a diffraction grating with horizontal slits, this "dispersion" or shifting of images of various colors away from the horizontal plane is just what would be expected.

3. Now suppose that a very small (pointlike) source of white light is used. Each wavelength in the beam will be diffracted through a different angle, so the scene can be viewed not only from position a but also from vantage points up toward b or down toward c. With different viewing positions, the image will be in different colors. If for example the eye is moved upward from a, the color of the image will change through the spectral colors, becoming red at some position such as b.

This effect is the origin of the name "rainbow hologram." Because the images change color but are not all seen superimposed, the effect is different from the color smears seen with ordinary transmission holograms. The colored images are now separated.

4. Finally, suppose a vertical filamentary white-light source is used. (The kind of long tungsten bulb that is sold in many places for use in refrigerators is satisfactory, but the glass must not be frosted.) It is fascinating to observe that with the eye near position a, the image is seen in shades of black and white. For example, a pair of black-and-white dice are seen as such. For other vertical viewing positions, colors will again appear in the image.

HOW RAINBOW HOLOGRAMS ARE MADE AND WORK

The first step in understanding how rainbow holograms work is to know how they are made. The process takes place in two stages. In the first stage a transmission "master" hologram of the scene is made in the usual way but specifically with a collimated, plane-wave reference beam. With such a reference beam and with objects fairly near the plate, a zone-plate hologram will be formed.

The second stage is illustrated in Figure 10.6. After it is processed, the master hologram is set up with an opaque screen against it. The screen has a horizontal slit with a vertical width of perhaps one centimeter or so in it. As noted earlier, if a small segment of a hologram is used, the image is seen as if you were looking through a small window. In the present case, the effect is as though you were looking through a slit window and would be narrow vertically. Already you can understand the retention of horizontal parallax and loss of vertical parallax.

The slit hologram forms a real and a virtual image. The *real* image is chosen for use. The real image is used as if it were a real collection of objects in making the "copy" hologram. The copy plate receives light rays from the real image, and it is also provided with a plane-wave reference beam. Thus a hologram of a holographic image is made.

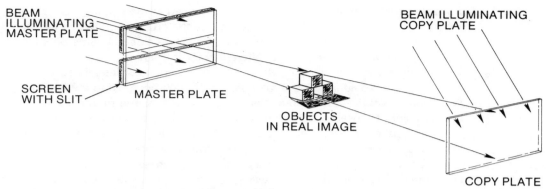

Figure 10.6. *Configuration for making a rainbow hologram.*

In reconstruction of the scene using the copy hologram and collimated light, the pseudoscopic image is used, so that the pseudoscopy of the image formed by the master hologram is reversed. The result is a normally orthoscopic view of the scene.

There is truth in saying that in viewing a rainbow hologram you are looking through a slit-window. If you have such a hologram and a laser you can carry out an interesting experiment. It requires two people in a fairly dark room. One looks into the hologram and moves his head up and down until he gets the best view of the scene. The other stands so that she can see the face of the other person. She will see a red streak—the real image of the slit—across the forehead or other part of the observer, with the red slit finally stretching across the observer's eyes when the desired orientation is reached.

Rainbow holograms tend to give bright images. While only the light falling on the slit in Figure 10.5 is used in forming the real image, once the copy hologram is made and then illuminated, *all* the light falling on it is put into the slitlike image.

In using a slit in this method there has been a cost—the loss of vertical parallax. This is not disturbing with rainbow holograms because we are so used to looking at the world with our two eyes more or less on a horizontal level, and because for the most part we move our heads horizontally and are therefore satisfied with only horizontal parallax.

Better rainbow holograms can be made by putting a cylindrical lens into the slit that is shown in Figure 10.5. An ordinary circular positive lens focuses to a point. A cylindrical lens focuses to a line. This focusing action can be used to advantage.

How can the color effects described earlier be understood? We offer a simple model. The interception of the zone plates by the slit causes the plate to have some properties of a horizontal diffraction grating. Diffraction of various colors in the incident light will take place as was discussed in Chapter 9. If monochromatic light of a color other than that used in making the plate were used, there would be a vertical displacement of the emergent beam. If two or three different colors were used, there would be two or three beams on each side of the horizontal plane, with a different color in each.

Finally there is the case in which a long filamentary white-light source is used. Huygens would approve of us thinking of each point along the filament as a source. Along the horizontal plane (with the eye at a in Figure 10.5) the various colors from all of the point-sources can be sufficiently in register to recombine as white light. Because of the dynamic range in the scene, some parts will be seen as bright and others as dimmer or even black. Absence of white light to various degrees is interpreted as grayness to various degrees. Therefore in a restricted region of viewing you may perceive the scene in shades of black and white. Outside that region the colors are not in register, rather as though a printing machine failed to place the red, green, and yellow printing faces properly on top of each other.

As was said, rainbow holograms were invented by Stephen Benton in 1969 (when he was with the Polaroid Corporation). Since then he has devised a modified method that improves the extent to which the images are in register, so that good "achromatic" images are produced, free of residual unwanted color effects. Rainbow holograms of this kind are not common, but you may be able to see one in a display in a museum of holography.

INTEGRAMS

Another hybrid type of hologram has the especially interesting property of being able to produce scenes in motion. We will call this an *integram,* but the names *integral hologram* and *stereoscopic hologram* are also used in the literature.

Our description of integrams starts with a consideration of stereoscopy, which has been made familiar to many millions of people through its use in movies and television to give an illusion of three-dimensional viewing. When you look at a scene, you see it from two different perspectives, one for each eye. Suppose that two photographs of the scene were taken, one from the perspective of the left eye and one from that of the right eye. When you look at the photographs in such a way that your left eye sees only the left perspective view and the right eye only the other, you react as though you were viewing the original scene. Actually, you are looking at only two-dimensional photographs, and while you are using binocular vision, you do not see the scene with true parallax. Three-dimensional movies and television work similarly, with the right and left views in different colors. Special viewing glasses permit each eye to see only one of the views.

Making an integram is a two-step process. The first step involves ordinary photography and results in a set of two-dimensional pictures of the scene. In the second step holograms are made of those pictures.

One way to carry out the first step is to have the scene on a central stage and to swing a camera around the stage, taking a sequence of pictures, as in Figure 10.7 (a). As an alternative, the stage can rotate in front of a fixed camera. In either case the frames in the filmstrip show the scene from different viewpoints, running partway or all the way around the scene. This can be done with a still-life scene or with action taking place. Notice especially that since this is ordinary photography, the filmstrip can be made outdoors in natural light or indoors with ordinary artificial lighting.

In the second step a transmission hologram is made of each of the photographs. These holograms are made on a length of emulsion, with each individual hologram in the form of a narrow vertical strip about 1 millimeter across. The height may be as much as 30 or 40 centimeters in a typical case.

For viewing the integram, the film is shaped cylindrically so that it looks like part (a) of Figure 10.8. The source of the readout light must be along the axis, shining down from above or up from below. You look at the virtual image from outside, as in part (b) of the figure. Monochromatic illumination is necessary to avoid the color smears that transmission holograms produce when white light is used.

Figure 10.7. (a) A way to make movie stills sequentially from all angles around a scene. (b) Eventually a strip of film is produced, with narrow vertical holograms side by side along it.

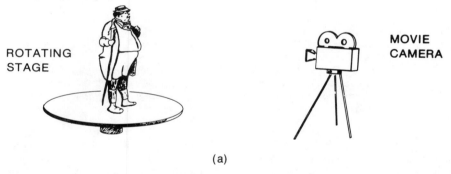

ROTATING STAGE

MOVIE CAMERA

(a)

STRIP OF FILM

(b)

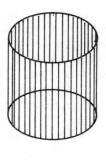

(a)

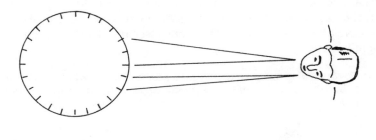

(b)

Figure 10.8. *(a) The vertical strips in a cylindrical hologram. (b) Overhead view showing an observer looking at the hologram, with his eyes seeing the strips from slightly different positions.*

Your eyes see different holographic images from different strips. This is similar to what happens in stereoscopy as described earlier, and as in that case you get an illusion of three-dimensionality. The virtual image is at the center of the cylinder and can be of any size. For example, it can be a life-size action portrait of someone, or it can be a group of miniature ballet dancers in action.

As you move around the cylinder or rotate it before you, your perspective of the scene changes. This does not happen continuously, because your eyes must switch from strip to strip, but in practice no discontinuity is noticed. The reason for this is interesting. If you have two ordinary photographs joined side by side, and move your line of sight across one, over the joint, and onto the other, you are of course aware of the discontinuity. But if you have two holograms side by side, as you cross the joint one eye will continue to see one of the images for a while after the other eye has picked up the second one; if the two images are only slightly different, you will not be aware of a sharp break. When ordinary motion-picture film is projected, a frame is shown for a short time, the projection is interrupted momentarily while a new frame moves into position, and this frame is shown for a short time. The old name "flickers" used to be an appropriate description. It is done more effectively today. If a truly holographic movie sequence could be made, this flicker action would not be needed. Instead, the film could be transported continuously in the projector.

The apparent three-dimensionality in the scene can be very convincing. For example, if there are two people in the scene, then from one viewing position person B may be hidden from view behind person A, but by moving to the right or left the observer can look past A and see B, as if looking at real persons. There is horizontal parallax, but it is not true holographic parallax, since each hologram gives an image of only a flat version of a scene. In this connection, notice that there is no vertical parallax!

The ways in which our eyes and brains interpret what we see are interesting, intricate, and little understood. You may be interested in referring to a very readable book with the self-descriptive title *Eye and Brain*,[3] by R. L. Gregory, in which such matters are discussed.

In one spectacular case of an integram, the entire earth was represented, as reported by Dr. Jeong.[4] The original pictures were gathered from an earth satellite, and these were converted into an integram. In a private communication Dr. Jeong has indicated that the result was not very good. Even so, the ingenuity in the idea is notable.

In the next section you will become acquainted with still another ingenious idea—the combination of rainbow hologram and integram techniques.

MULTIPLEX HOLOGRAMS

What today are called *multiplex holograms* first were made available by the Multiplex Company, after their development by Lloyd Cross in 1973. They are a cross (pun intended) between rainbow holograms and integrams. Rainbow holograms give still-life three-dimensional images but are viewable in white light. Integrams can give views of scenes in motion, with satisfying impressions of three-dimensionality, but they must be viewed with laser light. Lloyd Cross's accomplishment was to combine the two types, resulting in cylindrical films on which there are numerous vertical strip holograms made originally from ordinary movie frames. These can be viewed with an ordinary white-light source such as an unfrosted tungsten bulb, and present to the observer a view of a scene in motion. What the observer sees can be very striking, and at a holographic

[3]R. L. Gregory, *Eye and Brain*. (New York: McGraw-Hill Book Company, 1977).
[4]Dr. Tung H. Jeong, *A Study Guide on Holography*, 1975, available from Integraf, 745 North Waukegan Road, Lake Forest, Illinois 60046.

show a multiplex hologram is sure to draw forth oohs and ahs from the attendees. Technological advances often involve compromises, and in the case of multiplex holograms, the compromise is between impressive overall visual effects associated with the integram feature in them, and certain color effects associated with their rainbow hologram aspects.

Making a multiplex hologram is a two-step process. First a movie sequence is made, as with an ordinary integram. Then a vertical strip hologram of each frame in the movie is made, again as in the case of an integram, but now each of these individual holograms is made as a rainbow hologram, using a slit, with the slit vertical.

In order to be realistic in describing what an observer sees as he or she looks at a multiplex hologram, we will describe a popular one called "Kiss II," available from the Multiplex Company and other dealers. (See Appendix II.) The hologram is roughly 1½ feet long horizontally and 9 inches high vertically. It is shaped into a 120 degree arc of a cylinder and, as you can see by looking directly at it under room light, has hundreds of vertical strips in it, each about 1 millimeter wide. For a readout beam source, an unfrosted tungsten bulb is used. This is placed behind the cylindrical arc (from the viewer's side) and at a level a few inches below the bottom of the hologram itself. Since this is a transmission hologram, the observer looks into the film and sees the image on the other side, more or less at the axis of the cylinder. Direct light from the bulb can be blocked by opaque cardboard or other material.

Suppose you approach the cylinder to a distance of a few feet, with your eyes about the level of the vertical center of the hologram. You will see the head and shoulders of a very attractive woman, with one of her hands raised to about shoulder height. As you move to your right and circle around the cylinder (or rotate the cylinder before you), you see the woman follow your movements with her eyes, break into a smile, raise her hand, blow you a kiss, and then execute an entrancing wink.

Some multiplex holograms have been made in full 360 degree cylindrical form so that the viewer can see *continuous*—but repetitive—action in the scene simply by rotating the cylinder continuously. These are movies that involve holography.

All such holograms are tinged with color effects, since they are rainbow holograms. What color effects are seen depend on the level from which the image is viewed. In the case of "The Kiss," if you have your eyes about the level of the vertical center of the hologram you see most of the woman's hair, face, and upper torso in black, white, and shades of gray, with hints of colors appearing above and below. If you lower your head and look upward, blues or greens will predominate in the scene. If you raise your head and look downward, reds will predominate. The colors, their distributions vertically across the scene, and their effects on your appreciation of the scene vary as you move your head up and down. It is as though you were watching a live actress being lit by theatrical spotlights in a changing, and interesting, way.

These color effects are not related to the natural colors in the original scene. Rather, they are due to diffraction effects in the hologram. In fact, the movie sequence in the first step in making the hologram is in black and white, so the colors seen in the image must be holographic artifices.

COPYING HOLOGRAMS

Many transmission holograms, reflection holograms, projection holograms, rainbow holograms, multiplex holograms, and other kinds are available from dealers. Obviously good copies of hologram originals can be made inexpensively. How can multiple copies of a hologram be made? The simplest of all known methods is to cut a hologram into smaller pieces. This method depends on the property of redundancy: Each piece of a hologram is a hologram of the scene. Unfortunately, each piece of an originally large hologram offers only a restricted view, since a small window onto a scene restricts the viewing range, and the resolution provided by a piece is reduced. The resolution depends not only on the spatial frequency of the interference fringes, but also on the total number of fringes and, hence, on the size of the piece.

Better copies can be made by making holograms of the original "master" hologram. A method for copying reflection holograms is shown in Figure 10.9 (a). Light reflected back from the master hologram to the copy plate interferes with the light falling directly on the copy plate, and a copy hologram is formed. Excellent copies can be made in this way.

The equally simple arrangement in Figure 10.9 (b) can be used to make a copy of a transmission hologram. At the copy plate the reference beam is laser light that passes undisturbed through the master hologram, and the object beam is the diffracted beam from the master. This also can result in very good copies.

An embossing method that resembles that used in stamping out plastic copies of a master disc in the recording industry has also been used. If a master hologram has a surface relief structure, this can be metallized to make it durable enough to use in impressing the pattern into sheets of plastic. As noted earlier (Chapter 7), the optical information about a scene is better recorded in a volume hologram, so embossed surface holograms cannot give images fully faithful to the original scene. In the March 1984 issue of *National Geographic* such a copy hologram made a spectacular cover. Over ten million copy holograms were made.

Finally, a copy of a hologram can be made by making a transparency copy by ordinary photographic means. This converts a negative into a positive, but this does not ordinarily matter. If the original hologram is a collection of zone plates and such a transparency copy is made, each black ring in a zone plate in the original becomes clear in the copy, and each clear ring becomes black. However, a property of zone plates is that whether they are black-white-

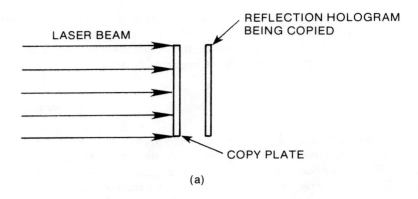

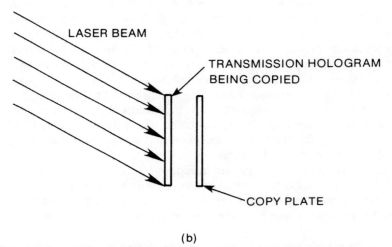

Figure 10.9. *(a) A method for copying a reflection hologram. (b) A method for copying a transmission hologram.*

black and so on, or white-black-white and so on, they produce the same images. Still, such a copy cannot be exactly like the original, and inevitably there must be some deterioration in the images that are reconstructed.

Some special techniques and applications

This chapter discusses some specialized topics in holography. Some examples of the special techniques included are: the use of sound waves instead of light waves; a kind of holography in which the objects provide their own reference beam as well as object beam; and interferometry. Among the applications are recognizing patterns, detecting minute motions or deformations of objects, viewing inside structures, and acquiring magnification so great that even some of the structure of atoms can be seen.

Most of the techniques included are actual, in that they have been accomplished, but technical obstacles have kept them from finding widespread and routine use in practice. The subject of special techniques and applications is so extensive that it will be continued in the next chapter.

HUGE MAGNIFICATION BY HOLOGRAPHY

The idea of achieving enormous magnification played an important part in the early development of holography. In 1947 Dr. Dennis Gabor was working in the research laboratory of the British Thomson-Houston Company, an electrical engineering firm, on the improvement of electron microscopes.[1] An electron microscope is a magnifying device that uses electron waves instead of ordinary light waves. Because the wavelength of electron waves can be many thousands of times less than for visible light, magnifications many times greater than in ordinary microscopy can be achieved. Electron microscopes are marvelous instruments that have contributed greatly to the advance of science, but they are rather conventional. Although they use electron waves instead of light waves, they function in a manner that is analogous to that used in ordinary microscopes, with special electric and magnetic field lenses to produce sharp and enlarged images.

Dr. Gabor's new idea was drastically different. In part it was to let coherent electron waves fall on the objects to produce an object beam, and to add to this a coherent reference beam so that interference fringes would result. Then as he says, "Let us make a hard positive record, so that it transmits only the maxima, and illuminate it with the reference source alone." This would reconstruct the image. Dr. Gabor adds, "This interference pattern I called a "hologram" from the Greek word *holos*—the whole—because it contained the whole of the information."[2]

In a second stage, the hologram would be reconstructed with visible light. As mentioned earlier (Chapter 5), the change in wavelength should yield a magnification. The magnification could in theory even be great enough to make individual atoms visible.

Dr. Gabor's proposal faced a fundamental bottleneck that still exists today— a practical source of coherent electron-wave beams does not exist. This is part of a more general problem—there also are no X-ray or gamma-ray lasers. Research is being carried out in these fields, with much of it classified because of military applications.[3]

In 1974 L. S. Bartell and C. L. Ritz, of the University of Michigan, reported making visually observable views of the electron clouds in atoms by means of a holographic microscope.[4] Electrons with energy of 40,000 electron volts were used. For these the wavelength is about 1/20 of an Angstrom, where the diameter of typical atoms is of the order of about 1 Angstrom. Diffraction and other effects set a theoretical upper limit of 0.08 Angstroms on the resolving power, but this is still roughly 100 times greater than for conventional electron microscopes.

---

[1]Dr. Dennis Gabor's account can be found in "Holography, 1948–1971," in *Science,* July 28, 1972, pages 299–313.

[2]Dr. Dennis Gabor, op cit., p. 300.

[3]See, for example, the following: G. Chapline and L. Wood, "X-Ray Lasers," *Physics Today,* June 1975, pages 40–48; G. C. Baldwin and R. V. Khokhlov, "Prospects for a Gamma-ray Laser," *Physics Today,* February 1975, pages 32–39; R. W. Waynant, "The Challenge of Short-wavelength Laser Research," *The Physics Teacher,* May 1976, pages 263–269.

[4]L. S. Bartell and C. L. Ritz, "Atomic Images by Electron-Wave Holography," *Science,* Vol. 185, September 27, 1974, pages 1163–5, and L. S. Bartell, "Images of Gas Atoms by Electron Holography," *Optik,* Vol. 43 (1975) No. 4, 373–390, and Vol. 43 (1975) No. 5, 403–418. Figure 11.1 showing an image of an argon atom is used with the permission of Dr. Bartell, who kindly let us reproduce a copy made from the hologram used in *Optik.*

Because of the lack of a source of coherent electron-wave beams, the usual scheme in which there is a coherent object beam and a separate coherent reference beam could not be used. Instead a method of holography was used which has long been known, though little used in optical holography. Chapter 8 talked about intermodulation noise. When light beams from various points on an illuminated object meet at the emulsion, they interfere with each other. Usually this self-interference competes with the desired interference of object light with the reference beam and is unwanted. In the method of Bartell and Ritz the electron waves fall on atoms in a gaseous sample. They are scattered weakly by the electrons in the atoms, but they are scattered strongly by the atomic nuclei. The basic idea is to use the strong waves from the nuclei as a reference beam, and the waves from the electrons as an object beam. To get a stationary interference pattern, coherence is still a requirement, but it happens that there is enough in the beams when this method is used. In short, each atom makes its own hologram!

Figure 11.1. Holographic micrograph of an argon atom. The magnification is more than 10^8. (This plate is from a special copy of the original hologram that was most kindly loaned by Dr. L. S. Bartell with permission to use it in this book.)

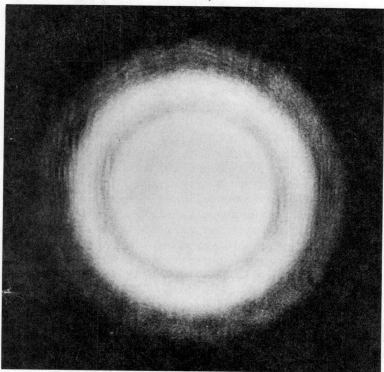

It may occur to you that, because of the random movements of the numerous atoms in the gas, the holograms self-made by individual atoms would in the ensemble produce a jumble. However, all of the atoms make their images centered on a common origin. All of the interference patterns fall on top of each other.

The interference pattern is recorded in special photographic emulsion that is sensitive to electron radiation. Then the scale of the record is changed by making enlargements in the darkroom. The final plate is viewed in red laser light.

Images of neon atoms and of argon atoms were obtained with magnification of 260 million diameters! A representative image is shown in Figure 11.1, page 155. Some of the rings are due to diffraction effects in the instrumentation, and interpretation of the image is not entirely straightforward. In their article Bartell and Ritz offer a quantum mechanical prediction of what the images should look like, and compare their real images with this. The agreement is convincing.

It would not be unreasonable to consider this a remarkable accomplishment, but significant further development has not yet been forthcoming. The method is beset with difficulties that include unwanted diffraction effects and even speckle. Also it is practical only for a severely limited class of specimens. Nevertheless, it forms an interesting and instructive case study for a student of holography.

OPTICAL MICROSCOPY USING HOLOGRAPHY

Magnification can also be achieved without having a change in wavelength between the making of a hologram and reconstructing it. Instead, optical laser light with the same wavelength can be used in both of the steps.

There are two ways in which this is done. In the simpler of these, the subject is holographed, and then the reconstructed image is enlarged by means of ordinary optical lenses. In the other method, magnification comes before, rather than after, the interference fringes are recorded. A laser beam is divided with a beam splitter. One of the beams goes to the object picking up optical information, and then goes through a microscope. When this beam emerges, it is mixed with the other beam, which serves as the reference beam. On reconstruction of the hologram, the previously enlarged image is produced.

In either case the most interesting question is what advantage is offered over ordinary "live" microscopy. As anyone who has used a microscope knows, the depth of field is very small. In order to look at various levels in a specimen, you must focus the microscope at various planes. If the specimen is alive and moving, focusing is impossible. However, suppose that the specimen is holographed with a short time exposure. Then the three-dimensional image can be examined at leisure at any depth, without disturbance from the motion.

STOP-ACTION HOLOGRAPHY AND HUMAN PORTRAITURE

The discussion in the preceding section leads us into a wider consideration of stop-action holography. This has as one of its aspects holographic portraiture of human beings, which deserves more attention than it has drawn.

In his Nobel address, Dr. Gabor asks us to imagine that a physicist is given the problem of determining the sizes of the droplets that issue from a jet nozzle at extremely high speeds, with the droplets so small that some are only a few wavelengths of visible light in size. "Certainly he would have thrown up his hands in despair! But all it takes now is to record a simple in-line hologram of the jet, with the plate at a safe distance, with a ruby laser pulse of 20 to 30 nanoseconds. One then looks at the real image (or one reverses the illuminating beam and makes a real image of the virtual one), one dives with a microscope into the three-dimensional image of the jet and focuses the particles, one after another."[5]

In the standard holographic setups described in Chapter 8, the apparatus must be kept rigid so as not to spoil the hologram. An obvious way around this difficulty is that indicated by Dr. Gabor in the case of the jet droplets: Use a pulse of light with duration so short that movement in the scene during exposure of the emulsion is not significant. Unfortunately, holographic emulsions have limited sensitivities. If they are exposed to light for very short times, the intensity of the light must be great enough to compensate.

Nevertheless, holographic three-dimensional portraits of people have been made. According to Wenyon, "in the U.S.A. it is now possible, though at considerable expense, to commission a portrait by a holographer."[6]

FULL-COLOR AND PSEUDOCOLOR HOLOGRAMS

It would be a great step forward in holography if it became possible to make holograms easily that would show scenes in their full natural colors, and that could be viewed simply with white-light sources. Even pseudocolor holograms that give good approximation to natural colors would be highly welcome. Nearly natural color holograms *can* be made; however, the techniques are difficult, and such a hologram is unusual. Ordinarily you can view one only at holography museums such as the Museum of Holography in New York City, or others in other large cities.

The basic idea is to superimpose three images, each in one of three primary colors. In color printing, for example, printed images in red, blue, and

---

[5]Dennis Gabor, op cit.
[6]Michael Wenyon, *Understanding Holography*, (New York: Arco Publishing Company, Inc., 1978), page 81.

yellow are superimposed. In color photography the picture is made using emulsion that contains layers that are individually sensitive to the primary colors. Again, in color television the same principle is used, but because this is a color-projection system rather than a color-substractive one, the primary colors are red, blue, and green.

In the case of color holography, a beam of coherent light in each of the colors red, blue, and green would be used in making the plate. The scene would be recorded in the form of a "red" hologram thanks to the red beam, and similarly in the form of a "blue" hologram and a "green" hologram. In order to reconstruct the image, readout beams of the same colors would be used, or white light, which contains the colors. The "red" hologram would act on the red component in the light to produce a red image and similarly the "blue" and "green" holograms would produce blue and green images. Assuming that the images are perfectly in register, a nearly natural-color scene would be perceived by the viewer.

That is the basic scheme, either for transmission holograms or for reflection holograms. Unfortunately, things are not that simple in reality. In earlier chapters you learned that when a transmission hologram is viewed in white light, color-smeared images are seen, and the situation would be worse if the plate contained three holograms—"red," "blue," and "green." Suppose that such a plate were illuminated with separate beams of red, blue, and green light. The "blue" hologram would give not only an image with the blue light, but also one with the red light and one with the green light. In all there would be nine images, and these would be of various sizes (because magnification depends on wavelength), and displaced from each other generally. To get good registration of *three* of these would require careful angular adjustments of the beams while making the hologram, and precise duplication of the conditions during readout. Also, emulsions have different sensitivities for different wavelengths, and in making the hologram considerable skill would be required with respect to exposure times and beam ratios. The method would also be very expensive, since the lasers required would cost many tens of thousands of dollars.

Efforts have been made to devise techniques to circumvent such problems, but it seems probable that reflection holograms offer better possibilities. Recall that a reflection hologram is a color filter. It selects out of a beam of white light a color that corresponds to that which was used in making the plate. Suppose that the hologram were threefold, as in the case of a transmission hologram, using red, blue, and green coherent beams. Detailed analysis of the effects of the component holograms on the colors of readout light indicate that red, blue, and green images could result. The plate would be a white-light viewable full-color hologram. One difficulty with this is the control of the incident directions of the three colored beams that is required to make the plate. Another is the additional (and not simple) problem of the varying sensitivity of the emulsion or other medium for different wavelengths. Experimenters in this field agree that still another problem exists, and it is the main one: the shrinkage of the medium during processing, which affects each readout color differently. Correcting for this is a challenge.

Holographers who have been able to make good full-color holograms are to be admired for their accomplishment. Good natural color holograms viewable in white light may become available someday. The goal is strongly desirable.

COMPUTER-GENERATED HOLOGRAMS

A hologram is a plate in which there are variations in darkness that form a meaningful pattern. Could such a pattern be *calculated,* recorded in a plate, and used to reconstruct an image? If so, holographic images could be produced of nonexistent objects, perhaps to create works of art, perhaps to study proposed industrial artifacts that might be too expensive to make in actual prototype form, or for other purposes.

It is possible to create holograms via computers. The mathematics of holography is well understood, and since modern computers are capable of carrying out voluminous calculations at high speed, it has been done. Of course it would take us beyond the purposes of this book to go into this subject in detail, but for those who like working with computers, have access to a small desk-top computer with a plotter, and have some facility in college-level mathematics, making a computer-generated hologram is possible. An article by J. S. Marsh and R. C. Smith is one of the best introductory guides to the subject.[7]

Whether you are that deeply interested or not, some interesting experimentation with a synthetic (computed) hologram is within your reach. The article referred to includes a printed computer-generated hologram, which can be duplicated photographically for viewing. Otherwise all you need is a laser and a simple lens. Instructions are given in the article.

INTERFEROMETRY

In the field of optics, interferometry is a technique that uses the interference between light beams to obtain useful results. It has been used to measure very small differences in wavelength, to detect very small movements, to measure lengths to great precision, and (in modern large radio telescope arrays) to resolve fine details in astronomical objects.

Holographic interferometry or *holometry* also is concerned with interference effects between two beams of light and also has varied applications. In

---

[7]J. S. Marsh and R. C. Smith, "Computer Holograms with a Desk-top Calculator," *American Journal of Physics,* Vol. 44, No. 8, August 1976. pages 774–776. Another resource of more general nature is by Maurice Francon (translated by G. M. Spruch), *Holography,* (New York: Academic Press, 1974).

one of its forms, it is known as *double-exposure* holometry. The object is illuminated with coherent light that reflects to a plate, which also receives a reference beam, as in the standard procedure for making a hologram. Then the plate is protected from light for a time, during which the object changes its position or its shape very slightly. Then a second exposure of the plate is made. After the plate is processed, it contains two slightly different holographic records.

For an explanation of what happens, we again turn to the words of Dr. Gabor: "We take a hologram of a body in state A. This means that we freeze in wave A by means of a reference beam. Now let us deform the body so that it assumes state B, and take a second hologram in the same emulsion with the same reference beam. Now the two waves A and B, which have been frozen in at different times, and which have never "seen" one another, will be revived simultaneously, and they interfere with one another."[8]

A sketch that shows the nature of what the interference pattern looks like in a particular case is given in Figure 11.2. The object is a bar of steel stressed by placing a weight on its top face. The fringe pattern reveals the strain pattern that results. In another application, the growth of a mushroom during a small time interval can be indicated holometrically. In this case the amount of growth may amount to only a fraction of a wavelength of light, and yet interferometry, unlike ordinary photography, can give unmistakable evidence of it. Furthermore, the information can be made quantitative, giving a numerical measure of the rate of growth.

Industrial applications include testing airplane parts for small welding defects and testing automobile tires. In the latter case, a hologram of the tire is made, then some warm air is blown on the tire, then a second exposure is made.

Figure 11.2. Image of a metal bar stressed by a weight on its upper face, as seen with double-exposure interferometry.

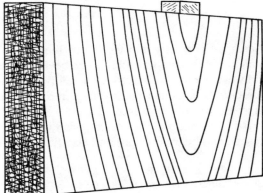

[8]Dennis Gabor, op cit.

Where a defect exists, the expansion of the tire at the location of the defect produces a readily detectable anomaly in the interference pattern.

In a related version of this technique, the real object is holographed but with only a single exposure. Then the reconstructed image is viewed superimposed on the original object itself. For example, the shaping of an optical surface such as that of a telescope mirror can be studied to a very fine degree of smallness. After the hologram is made and the mirror is subjected to some further grinding, the image and the mirror itself can be compared.

Still another interferometric method is a *dynamic* one. To convey the basic idea let's consider how it can be used to study the flight of a high-speed projectile in air. A laser beam passes through the air chamber to a plate, where a reference beam also shines. An exposure is made of the quiet air. When the projectile is passing through the chamber, a short high-power laser pulse is triggered. A second exposure is made. Changes in the air density due to compression by the projectile affect the parts of the beam when they arrive at the plate. Constructive and destructive interference regions are seen when the hologram is used to reconstruct the image.

Finally, there is *contour* interferometry. In one version, a hologram of the object is made using one wavelength of laser light, and then a second exposure on the same plate is made using a slightly different wavelength. The object is not changed in position or in shape between the exposures. Still, the plate contains two different interference patterns, and when it is viewed with one of the wavelengths, there are two readout beams that interfere with each other. The result is an image of the object with contour lines on it, like that in topographic maps.

Making double-exposure holographic interferograms is within the reach of students who have access to a simple laboratory, or to amateurs similarly equipped, but the dynamic and the contour methods require special apparatus which most experimenters do not have.

ACOUSTIC HOLOGRAPHY

Acoustic holography uses sound waves instead of light waves to create a hologram. As always in holography this requires coherent object and reference beams, which are brought together to produce an interference pattern. This step can be managed with relative ease, because excellent and simple sources of coherent sound are available. Ordinary cone speakers driven by electronic oscillators can be used at low audible frequencies up to about 20,000 Hertz. But frequencies ranging up to millions of Hertz also can be generated; for example, by using piezoelectric crystals driven by oscillators.

In one kind of acoustic holography the next steps after forming the interference pattern are to make a permanent recording of the interference pattern, and then to illuminate the record with a laser beam to create a *visual* image. Because the pattern is in the form of varying intensities of sound, rather than light, making the recording presents special problems, but it can be done. In another case, known as *real-time acoustic holography,* the pattern is not recorded, but rather is converted immediately into a visual image, with the advantage that a moving scene can be watched.

Let us stress that we are considering holography and resultant three-dimensional images. *Nonholographic* acoustic imaging is a much older technique that shows only two-dimensional images. For example, there are sonar devices that produce pictures on a screen after the manner of radar. Applications have included submarine detection and exploring for oil. As you know, acoustic imaging is also being used in medicine for examining unborn children and in detecting tumors.

This is a good point at which to explain the main advantage in using sound waves, whether in ordinary acoustic imaging or in holography. X-rays lose a substantial amount of energy as they travel through matter, and relatively little at an interface, which means they reflect poorly. The behavior of sound is distinctly different. It can travel great distances through matter without losing significant energy, and it reflects strongly at interfaces. For these reasons a physician can see a fetus in a womb, with its internal organs discernible clearly, or detect a brain tumor. This can be done safely. Exposure of the subject to the sound waves is not dangerous as would be the case with X-rays.

One advantage of acoustic holography over the older kinds of acoustic imaging has already been mentioned. Ordinary acoustic imaging responds only to the intensity of sound waves, as ordinary photographs respond only to the intensity of light waves. In acoustic holography, as in optical holography, both phase and amplitude information can be recorded and used to give a three-dimensional image. Another advantage is that acoustic imaging requires the use of acoustic lenses, which suffer from severe limitations, while holography can be lens-free. Still another is that the location of the sound detector relative to the object and the rest of the system is critical in ordinary imaging. In holography, the amplitude and phase information is present in any one plane, within wide limits, so that the system can be simpler in practice.

After these introductory remarks, let's turn to the question of how acoustic holography is actually done in practice. The principal method is *liquid surface acoustic holography.* In Figure 11.3 there is an object in a body of water. The object reflects a coherent sound wave from the object beam transmitter up toward the surface. A reference beam transmitter directs a similar coherent beam toward the same region of the surface. At the surface the interference produces a pattern of varying sound intensities. Where the sound intensity is high, the surface of the water is raised, and where the intensity is low, this does not happen as much. Thus the surface contains a pattern of mechanical deformation. The surface is the hologram.

A laser located above the surface shines a diverged beam onto the surface. The reflected light is photographed. The plate becomes a recording of the hologram and can be illuminated with laser light at leisure to reconstruct the image of the object.

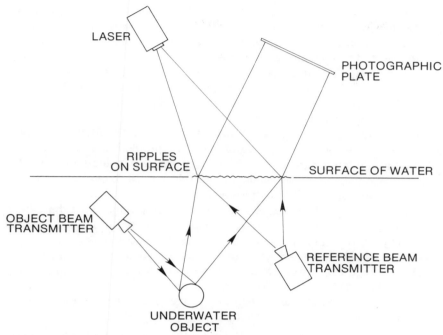

Figure 11.3. *One method used in liquid-surface acoustic holography.*

The sound used has a very high frequency, typically 5 million Hertz. One reason for this is that low frequency sound would yield a coarse ripple pattern on the surface, and the resolution in the ultimate image would be extremely poor. For example, with audible sound at 20,000 Hertz the wavelength in water is several centimeters but with ultrasound at 5 million Hertz the wavelength is less than a millimeter.

When the hologram is viewed with visible light, the change to a much shorter wavelength results in a magnification of less than one. Also, it puts the image a considerable distance away, so that a telescope may be needed to view it. Perhaps still worse, it greatly elongates the image, so that the desired advantage of realistic three-dimensionality is lost. Among other difficulties are that the surface of a body of water is subject to unwanted disturbances, and merely producing the interference ripples tends to produce streaming. A great deal of effort has gone into devising variations in the basic scheme to circumvent the difficulties, but we will not go into these matters in this brief introduction to the subject.[9] Instead, we will turn to consideration of some of the other major aspects of acoustic holography.

One of these is the real-time method. In this, the laser light reflected from

[9]Especially readable articles on acoustic holography are by A. F. Metherell, "Acoustical Holography," *Scientific American,* October 1969, pages 36–44, and by P. Greguss, "Acoustic Holography," *Physics Today,* October 1974, pages 42–49.

the surface is not captured in photographic emulsion, but instead is picked up by a television camera and the image is displayed on a special television screen. This avoids the time delay inherent in the other method and permits watching moving objects, but it requires an elaborate electronic system. A closer approach to the ideal viewing method would be a screen that could simply be placed in the reflected laser beam, much as a fluorescent screen is used in X-ray technology. Such a screen, consisting of a layer of a liquid crystal substance, has been used in experimental studies.

Could the sound waves travel in air? Eliminating the use of water as a medium would clearly be a great advantage. The problem with this idea is in part how the interference pattern can be recorded. In one method photographic film is exposed to light and placed in a fixing solution. While it is in the solution the sound waves fall on it. Where the sound intensity is high, the fixing process proceeds more rapidly than elsewhere. This and other methods that have been devised suffer from the severe limitation that the intensity of the sound must be *very* high, and that long time exposures are necessary.

The main alternative schemes use numerous microphones arranged in an array or a single microphone that scans the field in a raster. These transducers convert the sound signal they receive into electrical signals. Storage or immediate display of the record then becomes a matter of electronics. This also has both advantages and severe limitations.

MICROWAVE (RADAR) HOLOGRAPHY

Microwave radiation consists of electromagnetic waves, as does visible light, but it differs in that the wavelengths are much longer. Representative numerical values are 1/100 meter (1 centimeter) and 1/2 of one millionth of a meter respectively. Microwave radiation is much used today in communications and in cookery. Its use in imaging was developed during World War II in a form known as *radar.*

Radar imaging, like sonar (sound-wave imaging), resembles ordinary two-dimensional optical imaging in photography or television. It does not make use of both phase and amplitude information. *Radar holography,* also known as *coherent radar,* involves coherent object and reference beams. It also makes use of some novel principles and techniques.

In its best-known form, *side-looking radar,*[10] an airplane carries a generator of highly coherent microwave radiation, and the radiation is beamed out to the terrain being traversed. The reflected radiation is picked up by an antenna in the airplane. The received signal is mixed with a reference signal that is

---

[10]A readable and well-illustrated article on the subject is by H. Jensen, L. C. Graham, L. J. Porcello, and E. N. Leith, "Side-Looking Airborne Radar," *Scientific American,* October 1977, pages 84–95.

generated locally in the airborne apparatus. The mixing produces an interference pattern, and this is displayed on a screen, much as in television. The display is photographed to make a permanent hologram that can be illuminated with laser light to reconstruct an image of the terrain. The photographing can be done with motion picture film so that long stretches of terrain can be holographed.

A principal virtue in side-looking radar is that excellent visual images can be generated, revealing very fine details. This is due in part to the fact that the wavelengths used are short compared with features in the terrain, and in part because the radiation can have long coherence lengths. It is also due to another feature which is discussed below.

The radiation emitted from the airplane is in the form of plane waves, but as it reflects from individual points in the terrain, the object beams from those points are spherical waves. The reference beam generated in the airplane is a plane wave. Hence the hologram eventually formed is interpreted as a zone plate hologram, according to the ideas discussed in Chapter 9. A new feature is present, however, because the airplane *moved* during the time interval between the emission of the radiation and its reception back at the airplane after reflection from the ground. When the radiation is sent out, the emitting source and the receiving antenna are both in the airplane at one point along a path. When the reflected radiation is received, the point of reception is farther along the line of travel. While it is not obvious, though perhaps intuitively plausible, the effect is as though the signals were produced and received by a single straight antenna of considerable length. It is a general rule in optics that the larger the viewing instrument, the greater the resolution of detail that is possible. (This is why radio astronomers strive for ever larger arrays of detectors.)

Radar holography has some attractive attributes. As is so often true in advanced versions of holography, technical difficulties "come with the territory," and radar holography is not an exception. For one thing, the motion of the airplane introduces Doppler shifts in the frequencies of the radar signals, and these must be compensated for. The apparatus needed is also elaborate and expensive. Nevertheless, the technique is interesting and has important applications, including cartography, geological surveying, and military surveillance.

PATTERN RECOGNITION BY HOLOGRAPHY

This section covers *pattern recognition* and *pattern duplication,* with both being intimately related. The subject is covered here in rather broad strokes; further analysis is given in the next section.[11]

Pattern recognition can be explained by a simple example. A fingerprint (or even a fragment of a fingerprint) found at the scene of a crime is to be compared

---

[11]See also Yu. I. Ostrovsky, *Holography and Its Application,* translated by G. Leib. Mir Publishers, Moscow, USSR. Available from Academic Press, Inc., 111 Fifth Avenue, New York 1977.

with fingerprints on file. First a hologram of the fingerprint is made. This hologram is placed in the pattern-recognizer, along with a simple (nonholographic) transparency of a file fingerprint. A laser beam shines first through the transparency and then through the hologram. If the device recognizes the fingerprints as being the same, it signals by producing a bright dot on a screen.

The same method can be used not only to recognize but to *locate* some structure in an extended scene. For example, locating all the occurrences of some key word in a body of literature may be the objective. The device is presented with a hologram of the word and a transparency of a page of the text. Each time the device detects the word, it makes a dot on a screen, and the dots are positioned on the screen in a pattern that corresponds to the locations of the word in the text. A Biblical scholar might use this in preparing a concordance. It has been suggested that a federal agency might use it to search through voluminous statutes to find occurrences of such key words as "woman," "wife," "husband," and the like. Separating objects of a certain kind in aerial photographs and analyzing geophysical data have also been suggested.

Figure 11.4. *(a) The optical system used in pattern recognition. (b) On the screen the pattern of dots indicates where the letter s was found in the page of text.*

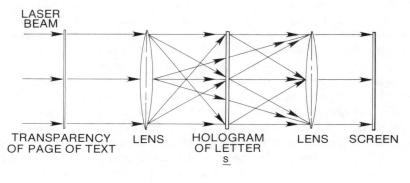

(a)

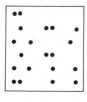

(b)

A similar device can *duplicate* an image. For example, such a device can be presented with a transparency of an intricate "mask" that is to be used in fabrication of complex electronic circuits, along with a hologram of a set of dots arranged in a raster. At its output, the device produces numerous duplications of the mask, side by side and row by row.

The actual process of *doing* pattern recognition and the apparatus required are quite simple. Suppose that the objective is to present the optical device with a page of printed text and to get an indication of all the locations in the page where the letter *s* appears. In the first step, a transparency is made that shows a clear letter *s* in an otherwise opaque background. Then a simple transmission hologram is made of this transparency. A transparency is also made of the page of text. Then the arrangement shown in Figure 11.4 is used. Laser light passes through the text transparency, through a lens, through the hologram of the *s,* through another lens, and onto a sreen.

What is seen on the screen is a set of dots. Each dot signifies that an *s* is recognized. The dots are positioned on the screen in a pattern that corresponds to the locations of the letter *s* in the text itself. Thus this simple device not only detects occurrences of the letter, but also specifies where they are.

The image *duplication* process is best thought of as a reversal of the process just described. Chapter 5 discussed the reversibility of light rays. In Figure 11.5 this principle is applied to image duplication. Light from the right first falls on a transparency in which there is a set of dots, arranged as in the output pattern on the screen in Figure 11.4. From there we trace the rays through a lens, through the hologram of the letter *s,* through another lens, and onto a screen, which in Figure 11.5 has replaced the transparency of the text in Figure 11.4. It follows from the reversibility principle that what is seen on this screen is a set of images of the letter *s.* An advantage of this is that the pattern of dots on the initial transparency can be arranged at will. If it is a regular line-by-line raster, the images can be made to fill the screen.

The question of how the "direct" process (Figure 11.4) works can be thought of in the following way. Before the laser beam falls on the transparency of the text, it obviously carries no optical information about the text. As the light passes through the transparency it picks up the information. It has been *modulated* by the transparency. When this modulated beam passes through the hologram of the letter *s,* further modulation takes place. This varies according to the different parts of the beam. Where there is agreement between the information carried by the beam and the holographic information, this second modulation is such that the *autocorrelation* is high. Parts of the beam carrying other information, such as letters *m* or *p,* are modulated differently. The autocorrelation is low. The optical information that has to do with occurrences of the letter *s* in he text has been reinforced, and other information has been repressed. After the second lens focuses the whole beam on the screen, there are regions of high intensity in the appropriate places.

Why must a *hologram* of the letter *s* be used instead of an ordinary

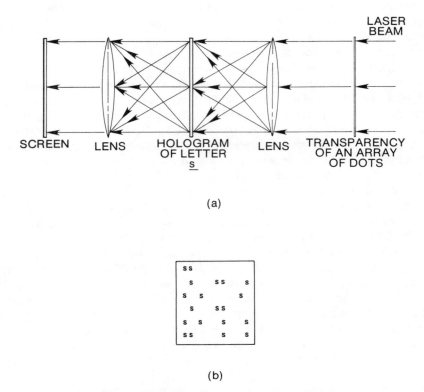

LASER
BEAM

SCREEN   LENS   HOLOGRAM   LENS   TRANSPARENCY
                  OF LETTER   OF AN ARRAY
                     s   OF DOTS

(a)

(b)

Figure 11.5. *(a) The optical system used for mulitple duplication of a symbol. (b) An array of images of the letter s as it might be produced.*

transparency? Nonholographic pattern recognizers have been studied, and some have become practical, as in some reading machines for the blind. Thus a hologram is not necessary. However, because so much more optical information is recorded in a hologram than in a simple transparency, the precision requirement on the optical system are much reduced, and even the hologram itself need not have very high resolution in some applications.

Aside from the question of advantages and disadvantages, in this book we regard the holographic technique as one that involves some interesting ideas for a student of holography. In this spirit, until now it would be fair to characterize the discussion as more description than explanation. It leaves open some important questions. What are the functions of the lenses in the system? Why are dots shown on the screen? What is the nature of the modulation of the laser beam by the hologram?

The hologram that is made for use in pattern recognition is a special kind called a *Fourier* hologram.[12] This special technique is not lensless holography. Let us first consider how a Fourier hologram is made, and then proceed to an explanation of it.

In Figure 11.6 a hologram is to be made of an object. The object is placed in the focal plane of a lens. The emulsion is placed an equal distance from the lens on the other side. A reference beam is brought in from the side to mix with the object beam in the emulsion. An interference pattern is recorded, and since this contains both phase and amplitude information about the object, it is a hologram, though not of the usual kind.

Now consider the various parts of Figure 11.7. In (a) light from the object falls on a screen that is so far away from the object that the light rays are essentially parallel. As you can easily verify by putting a piece of paper in the light from a distant light bulb, there will be illumination on the paper, but an image of the glass envelope and the bright filament will not be seen. In (b) a lens has been inserted with the screen in the focal plane of the lens. Thanks to the focusing action of the lens, an image will now be seen on the screen. With the same arrangement modified by placing the screen somewhere other than in the focal plane as in (c), an image will not be seen. It is "out of focus." Nevertheless the light on the screen carries optical information about the object, although it is not intelligible to us as an image. The nature of the information was worked out by the German physicist Ernst Abbé in the late 1800s by a

Figure 11.6. How a Fourier hologram of an object is made.

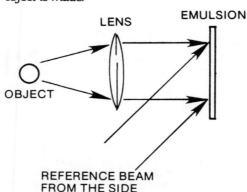

LENS   EMULSION

OBJECT

REFERENCE BEAM
FROM THE SIDE

---

[12]When a transmission hologram is made with the source of the reference beam and the object being holographed lying in the same plane, the hologram that results is known as a "lensless Fourier hologram." Actually, the hologram is *not* a Fourier transfer hologram.

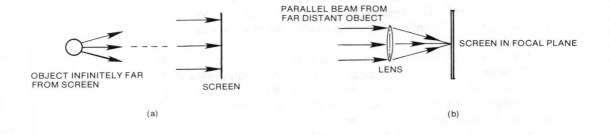

(a)   (b)

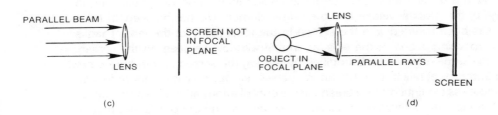

(c)   (d)

(e)

Figure 11.7. (a) Light from a distant object does not give an image on the screen. (b) A lens can be used to make an image on the screen. (c) If the screen is not in the focal plane of the lens it receives Fourier transformed light, but there is not a recognizable image. (d) Another case in which the Fourier transformed light at the screen does not give an image. (e) Two lenses in sequence can give an image.

mathematical technique called Fourier analysis. While we cannot go into this in any detail, we can recognize that the effect of the lens is understood, and adopt this terminology: The lens produces a *Fourier transform* of the object.

The nature of the Fourier transform depends on the configuration. If as in (c) the screen is moved, it receives light that is different in detail. In (d), the arrangement is also different, but there is a certain Fourier transform at the screen.

In (d) the *object* is in the focal plane of a lens, and the screen is somewhere off to the right. The screen does not show an image, but because the optical information needed is there, an image *can* be formed. In (e) this is accomplished by using a second lens, with the screen in the focal plane of this second lens. The second lens refocuses the light. In technical terminology, a second Fourier transformation undoes the first one.

When the Fourier hologram is made (Figure 11.6), the result is not an ordinary transmission hologram that can be reconstructed with a simple beam of laser light. Because a Fourier transform has been recorded, a lens must be used in the reconstruction in order to give a second Fourier transformation. Now you can return to Figure 11.4, which shows the setup for pattern recognition, and understand what is meant by saying that the hologram is a Fourier hologram, and what the purpose of the lens at the right is. Furthermore, you can understand that the lens at the left is producing a *Fourier transform* of the transparency at the hologram, and that in the hologram this transform is being compared with the optical information about another *Fourier transform*.

The combination of two Fourier images is known as *convolution* in the mathematical theory. How convolution works out with the arrangement in Figure 11.4 can be perceived with some depth of understanding without making use of the full theory. First suppose that the hologram is not present. If the transparency, the lenses, and the screen are positioned properly, you get a Fourier transform of a Fourier transform, and an image of the transparency is formed on the screen. As the light from the left lens passes over to the right lens, it contains a two-dimensional spread of spatial frequencies in any intermediate plane.

You can get a satisfying idea of what this means by analogy with what a diffraction grating does. A beam of light consisting of component colors is spread out by a grating so that a spectrum results. Short-wavelength (violet) light is dispersed relatively little, and long-wavelengths (red) light is dispersed more. Instead of speaking of wavelengths, let's say that light with high *temporal* frequency (violet) and light with low *temporal* frequency (red) are dispersed differently by the grating. When light from the transparency travels to the left lens (Figure 11.4), it carries optical information about gross features in the transparency, about fine-grained details, and about intermediate complexities. When monochromatic laser light is used, these are not differences in color (or temporal frequencies). They are differences in spatial frequencies.

Now for a critically important idea: You can place a filter in the position of the hologram shown in Figure 11.4, using a filter that blocks the low spatial frequencies, or the high spatial frequencies, or some band of frequencies. This is the basis of an old and well-known technique called *spatial filtering.*[13] For example, if the high spatial frequencies are blocked by the filter, then only a coarse image is produced.[14]

In the pattern recognition scheme, the filter is a hologram, which reinforces or attenuates high or low spatial frequencies according to the information stored in it. The hologram itself is a complicated set of patterns with varying spatial frequencies.

A letter *s* in the transparency results in a certain pattern of spatial frequencies in the light passed by the left lens. Convolution with optical information about a different letter in the hologram weakens all of the spatial frequencies, and convolution with a letter *s* strengthens the spatial frequencies, though not equally well at all frequencies. The low frequencies are the most enhanced. This already makes it plausible that the image is some kind of blob, since the higher spatial frequencies that distinguish between a human portrait, a Chinese character, and a letter *s* are suppressed. If you have to work with circular zone plates instead of linear gratings, it also becomes understandable that the recognition signal given by the optical system on the screen tends to be a *dot*. Notice that the fact that the recognition symbol is a mere dot does not mean that the pattern recognition process is a crude one. Instead, the more complicated the pattern to be recognized, the more reliable is the recognition. The more information it has to work with, the better the convolution process works.

[13]We suggest some articles on this subject. One is by Jearl Walker, "Simple Optical Experiments in Which Spatial Filtering Removes the 'Noise' from Pictures," *Scientific American,* November 1982, pages 194–205. Another is by Arthur Eisengraft, "A Closer Look at Diffraction: Experiments in Spatial Filtering," *The Physics Teacher,* April 1977, pages 199–211. Another is by George W. Stoke, Maurice Halioua, Venugopal Srinivasan, and Morimasa Shinoda, "Retrieval of Good Images from Accidentally Blurred Photographs," *Science,* July 25, 1975, pages 261–263.
[14]The lens-plus-pinhole spatial filter described in Chapter 8 is an instance of the opposite kind.

Holography in movies, television, and data storage

Chapter 11 gave a considerable range of holographic techniques and applications. This chapter continues with topics inherently of interest to a large portion of the public. Each could, once brought to a level of practical technology, be the basis for enormous industrial and commercial enterprises.

In addition, these subjects, involving as they do unusual and often ingenious tricks and twists, can hardly fail to be of technical interest to a student of holography.

HOLOGRAPHIC MOVIES

Millions of people have donned special spectacles with red and blue filters and watched 3-D movies. As explained in Chapter 10, these are based on stereoscopy. Each of the viewer's eyes receives two-dimensional images to look at, each showing the scene from a slightly different viewing angle. Even though there is no parallax in what the viewer sees, he or she gets an impression of depth in the scene.

On the other hand, the image produced by a *hologram* is a truly three-dimensional representation of the scene and can be very realistic. Furthermore, the viewer does not need special spectacles. It is no wonder that the dream of holographic movies has arisen. Have such movies actually been made? Yes, but severe technical problems have prevented their widespread use.

One of the earliest true holographic movies was made in the 1960s by Emmett Leith and his co-workers.[1] In Chapter 5, the idea of the multi-channel hologram was introduced. This is reminiscent of 360-degree multiplex holograms, which can be rotated before the viewer to give an action scene that lasts for a few seconds. With this scheme a small group of viewers can watch the cylindrical hologram simultaneously. However, while this is a kind of movie, it is not fully holographic.

In holographic movies there must be continuous action over a long period of time. At an early date it was proposed to accomplish this by recording holograms in a sequence of frames on a reel of film. This has actually been done.[2] In one case the film was seven centimeters wide (almost three inches) and the movie, which showed a goldfish swimming in a tank, lasted less than one minute. For viewing the movie, the film was illuminated with flashes of laser light, frame by frame, at the rate of about 30 per second. The brevity of this early movie is not significant, since clearly the basic idea could be used with longer reels. However, there remains the limitation that only one viewer at a time can watch.

This limitation can in principle be avoided by having the film pass from one viewing device to a second, then to a third, and so on, so that each person in the audience has his or her own projector. There are economic considerations as well as technical problems with this, and the method has not yet been made practical.

Finally, let's turn to a method for making and exhibiting holographic movies that has been used in the Soviet Union. Though the method still has limitations, it is interesting from a theoretical point of view. The following is based on an eye-witness account and a technical analysis by Dr. T. H. Jeong.[3]

The movies were shown in a dark theater setting to an audience of four at a time. The viewers looked at a screen about one meter square, several meters in front of them. Watching "almost as if the screen were a hole," they saw a girl holding a bouquet of flowers in front of her face. During the 45-second run the viewers could shift their viewpoints and see around and over the bouquet to look at the girl's face. The image was in shades of yellow, but in the dark room it gave an impression of being black and white.

In order to explain how the movie was made and exhibited, we will first

[1]Emmett Leith and Juris Upatnieks, *Journal of the Optical Society of America,* Vol. 53, 1963, page 1377, and Vol. 54, 1964, page 1295.
[2]A. D. Jacobson, V. Evtuhov, and J. K. Neeland, "Motion Picture Holography," *Applied Physics Letters,* Vol. 14, 1969, pages 120–122.
[3]T. H. Jeong, "Holography Now in the U.S.S.R.," *Optical Spectra,* April 1978, pages 41–43.

describe the basic ideas with some critical modifications omitted, and then explain those modifications. Light from a ruby laser is split. Some of the light reflects as an object beam from the scene onto the film, and some falls directly on the film as a reference beam. As the film is advanced frame by frame, the laser is pulsed synchronously and the holograms are made. To exhibit the sequence of holograms, the film is transported through the viewing system where it is illuminated synchronously.

This plan does not permit viewing by a number of persons in an audience. Clearly the modifications in this simple scheme are all-important. Our first step in examining these modifications will be a short excursion into some optical properties of elliptical mirrors.

Figure 12.1 (a) shows a circle. If this circle is reflecting on its inner side, then any ray of light from a source at the center will reflect back to the center, as in the case shown. In (b) a related curve is shown called an ellipse. An ellipse has two points, called _foci,_ which are such that a ray from a source at one focus reflects from the ellipse back to the other focus. (In the case of a circle, the two foci have merged into one.)

Now consider Figure 12.2 (a). The mirror is in the form of an arc of an ellipse. Near focus A laser light is illuminating a hologram to produce the real image at focus A. Light rays pass through this region and are reflected back to refocus at B. A viewer located as shown will be able to see the image.

Suppose that there were a set of similar elliptical mirrors present, with focus A in common, but turned so that their other foci B_1, B_2, and B_3 were at different locations. (See Figure 12.2 (b).) Each mirror would form an image at one of those three points. Assuming that the images are sufficiently separated, three different observers could be accommodated. With more than three mirrors, there could be more than three observers.

You may be wondering how three or more elliptical mirrors could actually be arranged as described. Fortunately, we don't have to do that. However, if the basic idea is to be put to use, something equivalent must be done. Recall that the interference of a plane wave and a spherical wave in an in-line arrangement

Figure 12.1. _(a) The circle represents a mirror silvered on its inner surface. Any light ray originating at the center reflects back to the center. (b) Now the mirror is elliptical. Any light ray originating at one focus of the ellipse reflects to the other focus._

(a)   (b)

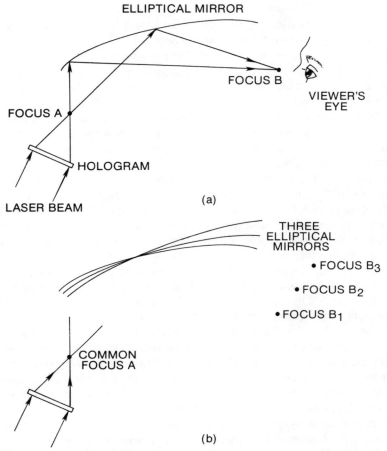

Figure 12.2. *(a) The real image is produced at focus A. The rays reflect from the elliptical mirror and the real image is reproduced again at focus B. (b) As before, a real image is produced at the focus A, which is common to the three mirrors. Three real images are reproduced at the different locations B_1, B_2, and B_3.*

produces circular fringes. In an off-axis arrangement the fringes become elliptical in form. If you think of the fringes as partially reflecting surfaces, as in the geometric model, then you can understand how elliptical mirrors can be achieved *in the hologram itself*. Furthermore, you can get sets of elliptical mirrors with different orientations of the sets in the plate. This can be done by a multiple-exposure technique. Each time there is the same reference beam, but the direction of incidence of light on the scene is changed, so that the relation between the object beam and the reference beam in the emulsion varies from one exposure to another.

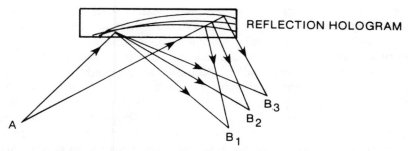

REFLECTION HOLOGRAM

Figure 12.3. A suitably made hologram contains in effect elliptical mirrors with different orientations. A light source at the common focus A reconstructs images viewable by separated persons at B_1, B_2, and B_3.

As stated earlier, the multiple recording of holograms requires that reflection holograms be made in thick emulsions in order to avoid overlap of images. In the Russian method, this is done, and with the spin-off advantage that the frames in the film are then white-light viewable. A pulsed laser is not needed in projection. To exhibit the movie, a white-light source is placed at A in Figure 12.3. Images are formed at B_1, B_2, and B_3.

While we have not attempted to exhaust the subject of holographic movies here, we have tried to give you an idea of what the subject is like and of what kinds of difficulties block its progress. The commercial importance of a practical scheme in which sizable audiences can see actors and scenes projected before them in natural color, as though they were really there, can hardly be exaggerated. But formidable problems exist. Some of these are associated with the holographic recording medium. For example, in the Russian method, you need a medium that will give numerous images of large size. If you are to retain the present movie system in which many theaters across a country can show one movie at the same time, that will require high quality and inexpensive copying of reels of film. The medium must also have high sensitivity so that actors need not be exposed to high power laser light and high resolving power . . . not to mention the problem of full-color images.

HOLOGRAPHIC TELEVISION

The possibility of holographic television is exciting. The commercial implications of achieving it are at least as great as in the case of holographic movies. It has been accomplished in laboratory studies, but it faces difficulties that must be overcome before it is put into practice.

The history of science and technology offers many instances in which developments that were unthought of before occurred, opening up vast fields of applications. It cannot be ruled out that if in the future holographic television is achieved in the practical sense, it will be on the basis of some ideas or devices

177

that are unknown today. For now we will have to suppose that holographic TV will work along the following lines. At the TV studio there are actors and props in a scene. Coherent light is used to illuminate the scene and the reflected light (object beam) falls on some kind of medium. A reference beam of the same kind of light also falls on the medium. As the action proceeds, holograms are made in a sequence of frames. Perhaps the medium will be photographic film, or if more exotic, still-like photographic film in that the interference fringes are recorded in it. The "medium" may also be a transducer that automatically converts the optical interference pattern into an electrical signal. (The TV camera in use today is a transducer of this kind, although it is used to convert non-holographic optical information from the scene into an electrical signal.) In the first case, the sequence of holographic frames in the reel of film will have to be converted by some special transducer in order to turn it into electrical signals.

In either case, the next step is to broadcast the TV signal to the antennas of receivers. Each receiving set will convert the electrical signal into a hologram frame by frame. In present-day TV each frame is "painted" on the screen line by line in a pattern known as a raster. In one holographic scheme, each hologram is painted line by line on the screen in similar fashion. What appears on the screen is not an image of the original scene, but a body of optical information that is unintelligible until an image is reconstructed with a suitable readout beam of light. In another scheme, the information in the electrical signal is collected internally in the set until a frame is completed, and then the whole hologram frame is displayed for reconstruction of the image.

There are numerous problems involved in this basic idea. One of them is especially troublesome; it is known as the "bandwidth" problem. It has to do with the amount of information that a hologram must contain if the recon-structed image is to be of good quality and with the time rate at which it is practical to transmit and receive the information via television.

In order to get an idea of the nature and magnitude of the problem, it is best to have some typical numbers to consider. A fringe density of only 100 lines per millimeter is too low for a good hologram, and 3000 lines per millimeter is the most attainable with ordinary photographic emulsion. Let us adopt the figure of 1000 lines per millimeter as a reasonable assumption. For a TV screen 10 centimeters by 10 centimeters in size (about 4 inches by 4 inches) there would be 100,000 lines horizontally and 100,000 lines vertically. The product—10 billion—is a measure of the amount of information per frame. In present-day TV the frames are shown at the rate of 30 per second. It follows that about 300 billion rudimentary pieces of information would have to be processed *per second.* (Such an astronomical number reminds one of the late Senator Everett Dirksen's saying that with a billion dollars here and a billion there, pretty soon you're talking real money.)

With current television technology, the rate at which information can be handled is about 100,000 times smaller than that. Even worse, for a screen of normal size, the practical rate is too small by a factor of many hundreds of thousands. It would take *days* to transmit one single picture!

Some attacks on the problem of holographic TV have been directed at reducing the amount of information per unit time required. To date the proposals fall considerably short of what is needed. Another idea is to increase greatly the rate of transmission of information (perhaps by a new optical fiber method). It may be that both reductions in the amount and the rate of transmission of information will eventually close the gap.

One way to reduce the amount of information is to transmit not entire holograms, but rather only horizontal strip holograms. The idea is to have the receiving set duplicate this strip line by line to fill the screen. If one entire frame consists of about 500 strips, as is the case in the currently used TV raster, the amount of information that must be transmitted is also reduced by a factor of 500. As mentioned earlier in connection with Rainbow and multiplex holograms, this would retain horizontal parallax while giving up vertical parallax. This also could be acceptable (unless you insist on viewing TV from a sidewise position).

Even so, the reduction factor is not great enough, so it has been proposed that the transmitted holograms be only small squares. At the receiver these would be replicated both horizontally and vertically. The reduction factor would then be 250,000 (500 times 500). Unfortunately this is still too little, and the horizontal parallax would be lost as well. With both vertical and horizontal parallax lost, is there any virtue left in the idea, which might cost vast amounts of money to develop and put into use? One answer is that stereoscopic views of the images on the screen could result, and it may be reasonable to think that such TV will come into being before true 3-D TV does.

Although we have been concentrating on the bandwidth problem, there are numerous other difficulties that will have to be faced. One already mentioned is the problem of the medium used in recording and reconstructing the holographic images. It must have high dynamic range (give good contrast in the images), and it must have high resolution. Also, it must be erasable and reusable, since it must show a frame for an instant and then show another, over and over, and its "inertia" (the time interval between display of one frame and the next) must be very small. Still another problem is that of producing a big picture.

After all this discussion, it may be surprising to learn that holograms have actually been transmitted, received, and reconstructed. We will describe here one instance as an example. A phototelegraphic channel between Moscow and Leningrad was used.[4] Line drawings, halftones, and still views of objects were transmitted. This is far from continuous action 3-D scenes. Still, the ability of holograms to reproduce fine detail resulted in about a *tenfold* improvement over the resolution

---

[4]Yu. I. Ostrovsky, *Holography and Its Application,* translated by G. Leib, Mir Publishers, Moscow, USSR. Available from Academic Press, Inc., 111 Fifth Avenue, New York 1977.

available with ordinary television. Dynamic range is also needed if pictures with good contrast are to be achieved. In this particular case there was only about two brightness gradations in the images, but the impression given was of satisfactory variations of tone in the images.

HOLOGRAPHIC STORAGE OF INFORMATION AND COMPUTER MEMORIES

It has become almost a truism to say that we are entering a predominantly data-handling age. Stupendous amounts of information must be recorded, recovered when needed, and altered quickly and accurately. With the advent of sophisticated microelectronic integrated circuit chips, a technological revolution in data-handling began, and its spreading influence continues.

How might holography be used in this field, how much has been accomplished so far, and what are the difficulties?

First let's distinguish between two kinds of information storage. In one kind pages of printed text, pictures, handwritten documents and the like are stored. This is done when books are shelved in a library and when correspondence is put into filing cabinets, for example, As is well known, a vast reduction in the volume needed can be effected by using microfilm.

In the other method, it is not literal replicas of text or pictures that are recorded, but rather strings of binary digits or "bits." A bit is simply a 0 or a 1. The manner in which text made up of letters, numbers and other symbols such as &, /, ? can be encoded in a string of bits is easily grasped. Suppose that 1111110 represents A, 1111111 represents B, 0000001 represents 1, 0000010 represents 2, and the other letters, numbers, and symbols are encoded in similar fashion. Then, for example, *YES?* might look something like 1110011110101111101001011101. Notice that pictorial information can be treated as well. To do this, the picture is broken up into minute pieces, somewhat as in halftone printing. The pieces, called *pixels* (picture elements), can be represented by bits. This is done when space vehicles transmit views of astronomical bodies back to earth.

Admittedly there is some inefficiency in this, because *YES?*—a word made up of only four ordinary symbols—becomes a twenty-eight-symbol "word" when coded as in our example. However, the scheme brings with it advantages that dwarf the disadvantage of inefficiency. One of the advantages is that it is not necessary to use the bits 0 and 1 *literally as such.* Instead merely a tiny spot can be used to represent 1 and the absence of a spot to represent 0. This is what is done when magnetic discs or tapes are used for computer data storage. There are minute magnetized regions and absences of magnetized regions in a string. When bits are so simple as to be merely spots or gaps in a record, they are simple indeed!

Furthermore, bits recorded in that fashion can be recorded with extremely high density. Small discs can hold hundreds of thousands of bits, and spools of tape can hold millions. Even computer memory integrated circuit chips can hold many tens of thousands of bits in an area considerably smaller than a fingernail.

Finally, such memory devices can be operated at very high rates, whether information is being recorded or read out. Operating speeds of several millions of Hertz are common.

After this preliminary excursion into binary data handling, let us return to the kind of information storage in which whole pages of text are to be recorded _as such_, and in which the viewer using the record will see the pages as they originally appeared in books, magazines, or newspapers, except perhaps not in full color.

The most basic way to do this is to use the multiple channel hologram technique, which was described earlier in this book. This requires making reflection holograms in a very thick emulsion, rotating the plate slightly between exposures. The text is read out by changing the orientation of the plate in the reconstructing beam of light. This is the basis for one of the early methods of making holographic movies discussed in this chapter. In that case and in the case of text recording, the number of holograms that can be recorded in a plate is not nearly as great as what can be accomplished by use of microfilm, with which more than 50,000 pages have been recorded in a square inch of film. That is more than 100 books of 500 pages each!

This does not make the prospects for holographic data storage seem promising. The situation appears even worse when we recognize that the three-dimensional realism in holographic images is not even put to use in recording flat pages of text. Why consider using holography at all? Part of the answer is that more optical information is stored in holograms than in ordinary representations of objects, and this fact might be turned to advantage. A special property of holograms is their redundancy. In dense storage on microfilm, a scratch or a speck of dust could destroy significant amounts of information. A similar flaw on a hologram does not prevent the whole image from being reconstructed.

Another part of an answer is that the shortcomings of the early method described above does not constitute the best that human ingenuity can devise. In fact, a better method has been devised and used. This _flying spot_ technique is illustrated in Figure 12.4. The plate contains an array of holograms, and a laser beam is directed at these in sequence. The reconstructed images are made available for direct viewing or for detection by automatic means, such as by photoelectric devices. The storage of holograms in an array with a density of several hundred pages per square inch has been achieved.

Now let's introduce another idea: recording information in _binary_ form. If a hologram is required to record a complex scene, substantial demands are

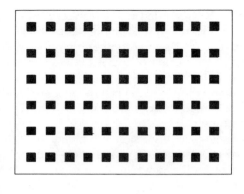

(a)

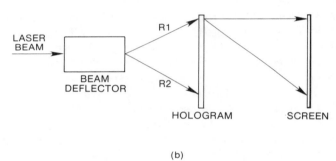

(b)

Figure 12.4. *(a) An array of holograms. Each small square represents the hologram of a page of original text. (b) A deflecting device sweeps the laser beam over the array, from side to side and from top to bottom. In a computer application, the screen would be replaced by transducers that would convert the optical signal into an electrical signal.*

made on it with respect to resolution and size. If only a pattern of dots is to be holographed, the requirements are greatly reduced. In holograms the spacing of the interference fringes is comparable with a wavelength of the light used; pages of binary information in the form of dots might be recorded on a similar scale, in principle. This suggests the possibility of compact storage of information, especially for use as computer memory.

Such a memory device should be capable of denser storage of bits than are magnetic systems, in which the magnetized spots must be relatively large. It is also possible to make the flying spot traverse the array at great speeds, so that fast reading of the data is possible.

However—and there seems always to be a "however" in advanced projects in holography—the dense recording and reading of binary information by optical but nonholographic means is also possible and has been accomplished. The clear superiority of a holographic method remains to be demonstrated. Also, the system gives what is called a *read-only memory,* (ROM), whether it is holographic or not. That means that the information, once recorded, cannot be altered selectively in places. Rather, the entire holographic plate must be remade incorporating the changes. The read-only feature is acceptable in some applications, such as compact recording of permanent libraries, but not in others, such as in banking operations, where constant updating of data is necessary.

What is needed to solve the latter problem is an easily erasable and reusable medium. (This need was also found in connection with the idea of holographic television.) As early as 1968 reseachers concerned with holographic data storage were able to achieve temporary storage in crystals.[5] Today other exotic media are known that can be erased selectively, but not as reliably and rapidly as is necessary.

In the computer field component parts and systems must come up to exceedingly high standards of performance and reliability. Optical computer memories have not yet been brought up to such standards.

[5]"Up to 1,000 Holograms Stored in One Crystal," *Bell Laboratories Record,* January 1969, pages 30–31.

How to make holograms: Part 1

This chapter and the following one discuss how to make holograms. Here (Part 1) our topics will be what your material requirements will be, how to go about satisfying them, and what the costs will be. In the next chapter (Part 2) our concern will be with actually making several kinds of holograms and with how you can go further in holography.

If you find yourself contemplating holography as a possible hands-on activity, we advise that you read these chapters through as a first step. This will give you an invaluable conception of the overall process. If you are a student in an institution that already has a holographic lab, you should nevertheless read these chapters to get an overview of holographic experimentation.

185

THE HOLOGRAPHIC TABLE

The surface on which you will work must be shake-proof, so that ideally it will not vibrate with an amplitude even as large as one-half wavelength of light. Given the severity of this requirement, it is remarkable how simple it is to meet the condition sufficiently and inexpensively.[1]

Figure 13.1 shows how a holographic table can be made. It consists of a slab of wood or metal, which rests on three or four partially inflated inner tubes, which in turn rest on a sturdy bench or table. The purpose of the tubes is to help keep the slab from shaking and to dampen out vibrations when they do occur. The working surface is the top of the slab. On it will be placed the laser, optical components, film holder, and objects in the scene. This kind of working surface is used with component mounts that are designed to sit directly on it, as chessmen sit on a chessboard. The slab can also be converted into a sandbox to accommodate component mounts and other items that are stuck down into sand to hold them in place. (See below.)

Any piece of flat wood that is heavy enough and large enough will serve. Plywood about 3/4 inch is easy to get and is good for the purpose. The main considerations about the dimensions are that the slab must be large enough to allow for spreading out of a holographic setup without cramping, but not so large as to encourage vibrations. A slab two or three feet wide and five or six feet long is suggested.

Figure 13.1. *A simple holographic table.*

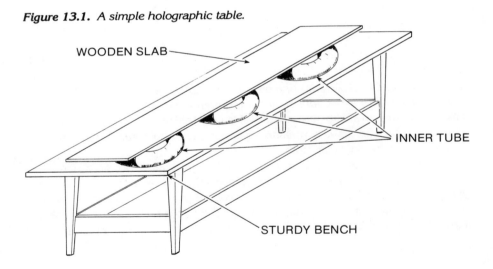

WOODEN SLAB

INNER TUBE

STURDY BENCH

---

[1]How to verify that the table is sufficiently immune to disturbances is described in Chapter 14.

If you do not have a table or bench that is the right size or solid enough, you can improvise. One way is to stack up concrete blocks to form legs and put another slab (as massive as you can manage) on the legs to form the top. Be sure that this top is solidly supported by the legs.

The inner tubes may be automobile or bicycle types, or smaller ones such as are sold in many stores for use in small vehicles. They should be inflated so that they are well filled out but still somewhat soft. A pressure about two-thirds of that recommended by the manufacturer probably will be about right. When the table is tested later (see Chapter 14) the tubes can be made softer or firmer as may be indicated.

Besides the chessman type of component mount, there is the stick-it-into-sand type previously mentioned. The optical component is held by glue or a clamp on top of a piece of tubing, or is squeezed into a slot in the tubing. The lower part of the tube is forced down into sand. It is then easily adjusted with respect to height and angles and is then held stably. If you decide to use such mounts, your holographic table can be the same as in the previous specifications but with side walls fastened to the slab to convert it into a sandbox. The side walls should be at least four inches high, preferably six inches high. The only care needed in building the sandbox is to make sure that the sides are strong enough and to avoid leakage of sand. Use white silica sand, the kind that is used in the floor-standing ashtrays. It can be bought at janitorial supply houses or from construction companies. Ordinary sand tends to settle slowly after it is disturbed, although we have used it successfully.

A sandbox perhaps 3-by-5 feet by 1/2 foot will hold such a quantity of sand that it will tend to resist shaking. On the other hand, it holds more sand than is needed and is unpleasant to remove. A simpler and cleaner method that we have adopted is to put the sand into plastic containers about 4 inches deep and some 4-by-10 inches or so in other dimensions, with flat bottoms. The containers sit directly on a flat slab. Plastic food containers that can be found in many stores work very well and are inexpensive. They help with stability, eliminate building the sandbox, and expedite handling.

THE HELIUM-NEON GAS LASER

Selecting a laser involves unfamiliar technical matters and a rather large expenditure. It might seem to the inexperienced experimenter to be very risky. Actually, you need only make sure that the laser you choose is suitable for holography, and compromise between the output power of the laser and the output power of your wallet.

The first rule is to deal with a reputable dealer, such as those listed in Appendix II. The dealer will tell you in his or her literature what applications the various models are suited for, will warranty his or her products, and will stand ready to help you. The dealer's descriptions will be easily read for the most part, having to do with shapes and dimensions and the like. One technical matter you should have a basic appreciation of is the *mode* of operation. Some lasers operate internally in such a way that the output beam shining on a screen across the room makes a complex spot. There may be a dark spot in the middle, or there may be a rosette pattern of lobes of brightness. On the contrary, what is wanted is a *uniphase* or *TEM$_{\infty}$* laser. Such a laser produces a spot that is round, free of structure, and yet fairly uniform in intensity, although it will be brightest near the center, fading off in its outer part. A second technical matter is the *polarization*. Polarized light consists of electromagnetic oscillations taking place in some plane, such as the vertical plane. Unpolarized light has no such distinguished direction of oscillation. A laser that produces either a polarized or an unpolarized beam can be used to make holograms, but the unpolarized (or random polarization) type is the better choice because it offers more output power for a given price.

The basic facts about output power are easily stated and understood. The importance of the power is that it determines the exposure time for a given setup. A 0.5 milliwatt laser can be used for simple setups, but it is at the lower limit of practicality. Exposure times may be thirty seconds, and a complicated multiple-beam setup may not be practical. At the other extreme is a 5 milliwatt laser. Exposure times from less than one second to a few seconds will result, depending on the setup. Safety is a significant consideration at such a power level. See "Laser Safety" in Appendix I.

A good all-around choice is a 0.8 or 1.0 milliwatt laser. All of the experiments in Chapter 14 can be carried out with such a laser, with exposure times running from three or four seconds to perhaps ten to fifteen seconds. Also, the intensity is adequate for viewing holographic images, giving very satisfactory brightness very safely. If you want to go to a higher power, a 3.0 to 5.0 milliwatt laser would be a good choice.

Output power and cost go hand in hand. Different makes of lasers of a given power vary widely in price, as is true of TV sets, automobiles, and most things, and prices change rapidly. The following table gives a rough idea of costs.

OUTPUT POWER (IN MILLIWATTS)	PRICE RANGE (IN DOLLARS)
0.5	300–350
0.8	350–500
2.0	450–800
3.0	500–900
5.0	600–1000

Diverging Lenses

In every setup at least one lens is needed to spread out the laser beam, and two are needed in a two-beam method. A diameter of about 1 centimeter is convenient—small enough for mounting on small stands, large enough to be hit easily in the center with the laser beam—without danger of light hitting the edges, which would spoil the beam for holography. The focal length is determined by the requirement that the beam will spread out enough to cover a scene or a plate in a distance compatible with the size of the table. A focal length of about 5 millimeters should serve well. If need be, two contiguous lenses can be used to get more rapid expansion. Either positive (converging) or negative (diverging) lenses can be used. A positive lens first brings the beam down to a focal point, after which the beam diverges, while a negative lens produces immediate divergence. Good lenses can be bought as inexpensively as $5 to $10 for a pair. Some dealers are listed in Appendix II.

Beam Splitter

A beam splitter can be very fancy, with partially silvered faces and precision manufacture and cost hundreds of dollars. Perfectly acceptable ones can be bought at prices less than $10 from Edmund Scientific and other dealers, and this is clearly the way to go if you choose to buy one or two. However, a piece of glass 1-by-2 inches or so will serve very nicely. The thickness should be 1/4 inch or so (double-weight window glass) so that secondary beams can be blocked off easily. (See Chapter 8.)

FRONT SURFACE MIRRORS

You need to have two *front surface* mirrors for some of the methods discussed in Chapter 14. Each should be at least 3-by-3 inches in size. (Smaller sizes have to be placed near the diverging lenses and are not versatile enough.) Fortunately, these are inexpensive items, with somewhat larger ones costing about $10 each. It is essential to check your mirrors for cleanliness. Reflect a diverged laser beam onto a distant screen and inspect the spot for blotches. If cleaning is necessary, you must remember that the silvering is on the front face and so is vulnerable to scratching. To clean, run alcohol over the face, check again, and if necessary, swab lightly with facial tissue. You can also wash in water with a little detergent in it, if this is done gently.

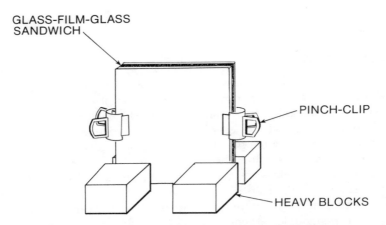

GLASS-FILM-GLASS
SANDWICH

PINCH-CLIP

HEAVY BLOCKS

Figure 13.2. *In the film holder the film is squeezed between two glass plates by two clips. The assembly can be held upright on the table top by some heavy weights.*

FILM HOLDER

If you are going to use flexible film, you will need a film holder and a way to stand the holder and film upright on the table. This is so easily fabricated that we will assume you are going to assemble your own. See Figure 13.2. The film is squeezed between two pieces of plate glass, each about 4-by-5 inches in size or less. The squeezing is provided by two large pinch-type paper clips. If this holder is to sit on a flat table, you can brace it with heavy blocks of metal as shown in the drawing. Such a film holder is shown in Figure 1.4. If you use a sandbox holographic table, simply stick the assembly down into the sand. In this case, the glass plates should be large enough to provide enough area above the surface of the sand to accommodate your film pieces.

If you use plates rather than film, and a flat top table, only the weights are needed to hold the plate in position. With a sandbox table, some of the plate must be inserted into the sand. Because film is so much less expensive than plates, you probably will not be faced with this problem.

SHUTTER AND BEAM BLOCKERS

Exposure times will be long (from a few to several seconds). Therefore they can be controlled manually and with a very primitive shutter. We have used an empty cigar box sitting on one of its smaller faces directly on the flat table top. A block of wood would be even better. With a sandbox system, a piece of black cardboard shoved down into the sand is all that is needed. Notice that the shutter need not be shake-free.

When you have set up components for making a hologram, it is essential that you check for the possible presence of light beams where they are not wanted, and block them off when they are found. For this purpose also blocks of wood, cigar boxes, or pieces of cardboard will serve.

SPATIAL FILTER

A spatial filter is a device that is placed into the laser beam where the beam emerges from the laser, and converts a blotchy, "noisy" beam into a diverging beam with vastly improved freedom from blemishes. The improvement can be spectacular to behold. Chapter 8 discussed how a spatial filter works. A good one can be purchased for a little more than $100. If you are just starting out and must economize as much as possible, you should dispense with a spatial filter and consider acquiring one later.

In principle, it is not hard to make a spatial filter. It is at bottom a small lens, an opaque screen with a hole in it, and a means for making fine adjustments in the parts of the device and in the orientation of the hole in the laser beam. In practice, this is a challenging undertaking. If you choose to take this route, your best guide is the *Holography Handbook* by Fred Unterseher and co-authors, referred to in Appendix II.

COMPONENT MOUNTS

Lenses, mirrors, and beam splitters must each be held on some kind of stand, resistant to vibrations, and yet easily adjusted in the horizontal plane, easily adjusted up and down, and easily changed in its angular orientation. There are two basic schemes. In one, the mounts sit on the plane surface of the slab. In the sandbox type, the mounts are stuck down into the sand.

In the case of a flattop slab, laboratory-style stands and clamps can be used. In a school where there is a supply of such things on hand, this is feasible. For the home experimenter, components mounts can be easily made. For example, blocks of wood can be used with spring clips atop them to hold the components. Another system uses flat pieces of iron that lie on the slab and magnets at the bottoms of the mounts to hold them to the iron.

The sandbox system is especially popular because of the simplicity of the mounts required and the great flexibility in adjusting the mounts in making a holographic setup. Each mount need be no more than a length of tubing with the lens or other item cemented to its top. PVC tubing is very suitable for the purpose. Such mounts are easily made by even all-thumbs people.

DARKROOM EQUIPMENT

Let us first describe the steps you would go through in processing film you have just exposed. The procedure described is by no means the only one possible, but it is one of the simplest.

Imagine yourself in a room from which outside light is excluded by closing the window curtains, shutting the door, blocking off cracks, and ventilating ducts or other places where light could get in, but in which you can still see well enough to find your way around, and locate equipment, thanks to a dim green safelight. You go to a table where there is a series of trays, side by side. The film is put into a developing solution in the first tray, and because of the dim illumination in the "dark" room you can watch the film. After one or two minutes, the film turns dark. Using cheap plastic tongs or a wooden clip-type clothespin, you take the film out of the developer and look through it at the safelight. The film is fully developed if it is so black that you can just see the light through it, if the hologram is of the transmission type (about 20 percent transmission), or if you can estimate that the transmission is 50 percent or more for a reflection hologram.

The film is then put into water in the next tray and sloshed around to rinse it well. After a minute of rinsing, the film goes into the next tray for bleaching.[2] This is always an interesting stage to watch. After a minute or so, you see the blackness vanish, usually starting at one corner and progressing rapidly across the whole film. When the film is transparent, it is all right to turn on the room lights.

The next step is another washing. Because the purpose is to remove chemical residues thoroughly, this washing should be done carefully for several minutes. It is best if you can manage it with water from a tap running into the tray with the overflow going into a sink, although this is not necessary if the tray holds a good volume of water.

You now almost have a viewable hologram, but there is one last step to go through: drying. The simplest way to dry the film is to hang it somewhere, using a clip at one corner. Because you want to view your hologram as soon as possible, a more satisfactory procedure is to use an electric hair dryer of the blower type to dry the film. To avoid water spots on the hologram, a washing in Kodak Photo Flo solution before drying is desirable.

Now let's return to our main subject in this section—the darkroom facility and a shopping list of materials. It should be clear that a darkroom that would satisfy a nonholographic photographer is not necessary. Holographic film is insensitive to light, compared with photographic film, and some stray light over short periods of time will not matter. Also, most holographic film for red laser light is insensitive to green light, making the blessing offered by use of a dim green safelight possible. Notice that the processing facility need not be in the

---

[2]In some processes, the film goes through a stop-bath stage before bleaching.

same room as your holographic table. If you put the exposed film into a light-tight container, you can take it elsewhere for processing. The requirements on the darkroom are so relaxed for holographic purposes that you can even use a room in which there is a fair amount of light, especially if you protect the film well before starting and keep the *developing* tray covered. Notice also that the water tap and sink do not have to be in the darkroom. It is a rare person who cannot manage to concoct adequate processing facilities.

Our account of a typical processing of exposed film should make it clear what the basic necessities are. The safelight can be a green photographic bulb, a low wattage bulb, or even a flashlight covered with transparent green plastic. The trays can be photographic processing trays, heat-resistant glass cooking trays such as those used for making shallow casseroles, or even beakers or jars (if flexible film reasonably small in size is used). The plastic tongs are an inexpensive item available at any photography shop. A portable hair drier is not necessary, though useful for drying film quickly. While some chemicals used in processing recipes are highly toxic, we assume that you will avoid these and use safer methods. Even so, if only Kodak D-19 developer and an innocuous bleach are used, you may want to wear rubber gloves during processing.

FILM AND CHEMICALS

The subject of film for making holograms is an extensive and highly technical one. Fortunately if you are starting to experiment in holography, there is no need to be apprehensive about this, because there is one preeminently good choice. Use Agfa-Gevaert 8E75NAH, which is also known as 8E75HD. Film is preferable over plates because it is much cheaper, and because you can easily cut it to size and shape it to suit your needs. This particular film is best because it has high resolution, can be developed and bleached by a simple procedure that does not involve toxic chemicals, and is readily available. It has low sensitivity[3] (about 200 erg/square centimeter) compared with photographic film, yet exposure times with it are short, and the low sensitivity works against accidental fogging due to room light.

"NAH" means "non-antihalation." With silver halide emulsions, the development process may spread the originally exposed grain regions in a halolike effect. Antihilation, or AH, film has a bluish backing on one side to prevent this. AH film is not to be used for making a reflection hologram because the reference beam and object beam must reach the film from opposite sides. NAH film can be used for both reflection and transmission holograms and is the best single choice. The "HD" in 8E75HD means "high density." Sources of supply for 8E75 film are listed in Appendix II.

If you are a photography enthusiast or an experimenter with a special

---

[3]The SAE number is about 0.03.

interest in holographic media, you can easily try other films, or even more exotic media. The best source of information about this is the *Holography Handbook* by Unterseher and co-authors, referred to in Appendix II.

The development, fixing, and bleaching process is not discussed fully here because it is an extensive and highly ramified subject. It is also unnecessary. When you are starting out, all you need is one trouble-free method. Furthermore, you can expect to get processing instructions with the film you buy.

The following describes the procedures we have adopted and have found to be fully satisfactory. These procedures came to us originally from INTEGRAF, listed as a dealer in Appendix II. We recommend them, but you may get other instructions with your film if it is other than 8E75, and you should follow them.

For transmission holograms, (1) develop in Kodak D-19 for five minutes, (2) use Kodak stop bath for twenty seconds, (3) use Kodak rapid fix without the hardener for two minutes, (4) wash for five minutes, (5) bleach until clear, and (6) wash for ten minutes. All the chemicals are available at any photography shop except for the bleach chemicals. The bleaching solution is 150 grams of ferric nitrate and 33 grams of potassium bromide mixed in one liter of water.

For reflection holograms, (1) develop in Kodak K-19 for five minutes, (2) wash for five minutes, (3) bleach until transparent, and (4) wash for ten minutes. The bleach solution is 2 grams of potassium dichromate, 30 grams of potassium bromide, and 2 cubic centimeters of concentrated sulfuric acid, in one liter of water. (Use caution with the acid.)

The chemicals needed can be obtained at photographic supply houses or chemical supply houses. Try the chemistry department at a nearby college or university. Your pharmacist also may be able to help. However, they are not esoteric or scarce chemicals.

Another processing procedure has been developed recently that we recommend. It gives brilliant holograms, and in the case of reflection holograms, so much so that they can be called spectacular. Objects holographed can be seen with great clarity under almost any light source with striking realism. On flipping the film over, the image is almost as sharp, and the pseudoscopy is impressive.

The first step in this "pyrochrome" process is to prepare two solutions. One is simply 10 grams of pyrogallol per liter of water. The other is 60 grams of anhydrous sodium carbonate per liter of water. Just before you are ready to process an exposed film, mix equal parts of the two solutions together. (This is similar to the mixing of the two components of epoxy just before gluing two things together.) The reason for this on-the-spot mixing is that the *mixture* becomes bad quickly, so it cannot be stored. Discard any that is left over.

Next you will need the bleaching solution. This consists of 4 grams of potassium dichromate and 4 milliliters of concentrated sulphuric acid in 1 liter of water. Be careful with the acid!

Once the solutions are at hand, the method is simple: (1) mix equal parts of the first two solutions together as described above, put the film in the mixture, and develop for one minute; (2) wash for three minutes; (3) bleach until clear, using the potassium dichromate solution; (4) wash for three minutes; (5) wash in Kodak Photo Flo for two minutes; and (6) hang the film up to dry.

To conclude this section let's consider prices. A roll of 8E75 film 5 feet long by 10 inches wide costs about $40. This will make about 100 pieces about 2-by-3 inches in size—small but adequate for many experimental purposes, or about 30 pieces if the size is more like 4-by-5 inches, which is too large for most purposes. (On the other hand, plates 4-by-5 inches in size cost several dollars each.)

THE BUY OR BUILD QUESTION

Clearly you will have to build a holographic table. You also will have to buy lenses, film, and some other items. If you choose to build as many items as you can, you have to do so without any experience to guide you in either the design or construction. If you buy equipment piecemeal, you have to get catalogs from dealers and make decisions, again without experience to guide you. One excellent way to get started is to build your table, buy such minor items as trays and a squeegee locally, and otherwise equip yourself by buying a holographic kit. The kit will enable you to begin experimenting at once, give you equipment that you can count on to serve well, provide experience pleasurably gained, and all this at a cost so low that the wisdom of cutting corners by making things or buying them piecemeal becomes questionable.

An example of a holographic kit is the Metrologic Sandbox Holography Lab. This contains two diverging lenses, two front surface mirrors, a beam splitter with each of the foregoing on a carrier, a film holder consisting of two glass plates and two clamps, 72 sheets of film, developer, bleach; a safelight filter, shutter cards; apparatus for making 360-degree holograms that comprise a jar, base mirror, pedestal, and a subject; a good set of demonstration holograms, and a helpful instruction book. The current price is $130.

THE BOTTOM LINE

It has been said correctly that setting up a holography lab can be done at a cost comparable with what is required to get started in photography, but a difference is that while photography amateurs seem always to have other expensive desiderata in mind, once you have a basic lab established, you can go on to considerably more sophisticated levels at the minor cost of adding lenses and mirrors and the like.

In the table that follows we suggest in a time of volatile market prices how low the cost can be. We assume that you have a room to house the table, such as a screened-off corner of a basement, that at least a modest darkroom facility exists, and that a kit is to be purchased.

Approximately 1 milliwatt laser	$400
Holography kit	130
Materials for table (slab of wood, sideboards or sand containers, sand, inner tubes)	20
Miscellaneous (darkroom trays, squeegee, tongs)	20
Spatial filter	130

The total without the filter is $570; with the filter it is $700. In any event, the laser accounts for most of the total. If your interest in scientific experimentation and in optics in particular extends beyond holography, a significant consideration is that a laser opens up to you a scientific field that is rich in fascinating phenomena, including properties of lasers, interference and diffraction effects, voice modulation of a laser beam, laser light shows, and much more.

How to make holograms: Part 2

With the material requirements met, you are ready to test your holographic table and then begin to make holograms. This chapter explains the Michelson interferometric test, which has a forbiddingly technical name but is very simple and beautiful, and then discusses several ways to make holograms. These are the simplest one-beam and two-beam methods for transmission and reflection holograms. Each is interesting to work with, and each can produce a good hologram of its type. Finally, you will learn how to go further in holography—a field rich in possibilities for the experimenter.

TESTING THE TABLE

There are two reasons why it is essential that you make the Michelson interferometric test before you begin to make holograms. One is that you must be sure that the table does not shake under normal conditions, in spite of such disturbances in the environment as movement of people in the building, air

197

currents, and outside traffic. Holograms cannot be made on a shaking table. The other is that you need to know how rapidly the table settles down after it *has* been significantly disturbed. Setting up the interferometer as your first experiment is an excellent way to become familiar with your apparatus.

Figure 14.1. shows the geometry of the setup and the apparatus needed. Some of the laser beam reflects from the first surface of the beam splitter and some passes through. These beams from the splitter travel down to front-surface mirror A and to the right to front-surface mirror B. Ray A returns from its mirror to the splitter, passes through onto a screen, which can be a wall of the room, a piece of paper, or cardboard six feet or more away. Ray B returns from *its* mirror, reflects from the splitter, and goes to the screen. Each of these beams will show up as a spot of light on the screen. Actually, you will see more than two spots because there are reflections from both of the faces of the beam splitter. Work with the brightest spot from each mirror and ignore the others.

The beam splitter should be at an angle of 45 degrees to the laser beam. Beams A and B from the splitter will be at right angles to each other when this is achieved. If you "eyeball" this reasonably well, the spots on the screen should be near each other, and then you can make fine adjustments to bring them closer together until they are *superimposed.* At the same time the distances from the beam splitter to the mirrors should be equal or nearly so. A simple way to check this is to use a piece of string to compare the distances. If you change one of the distances, the interference pattern will be affected. The best pattern results with equal distances.

When the spots are superimposed, the two coherent beams will produce interference fringes in a small region. You need to make the fringes easily visible. To do this, insert a diverging lens to spread the light out on the screen, as shown in (b) of Figure 14.1. Because the interfering wavefronts are spherical, the fringes you see will be curved arcs, as illustrated in Figure 14.2. Even after you have seen this effect many times, it will remain fascinating.

If the interferometer is set up correctly, getting visible fringes is guaranteed provided nothing in the system shakes. If you do not see the fringes yet are sure that the slab is stable, the most likely source of trouble is the shaking of *components* on the table. Make sure each optical component is solid on its mount and that each mount is solidly seated on the table or held securely in the sand. If it appears that the table itself is shaking, try changing the inflation of the tires, placing them in different positions, and perhaps putting weights on the slab to increase its mass.

When you have a good fringe pattern, visible and at most perhaps shaking very little, tap the table. The fringes will disappear at once, leaving only a uniform bright region. In a short time—perhaps a second or two—the fringes should be coming back into view and after a few more seconds the pattern should be stationary again. Repeat this several times to get familiar with the *time scale* on which the quick disappearance and slow reappearance takes place, so that you have it in mind later when you no longer have the interferometer set up.

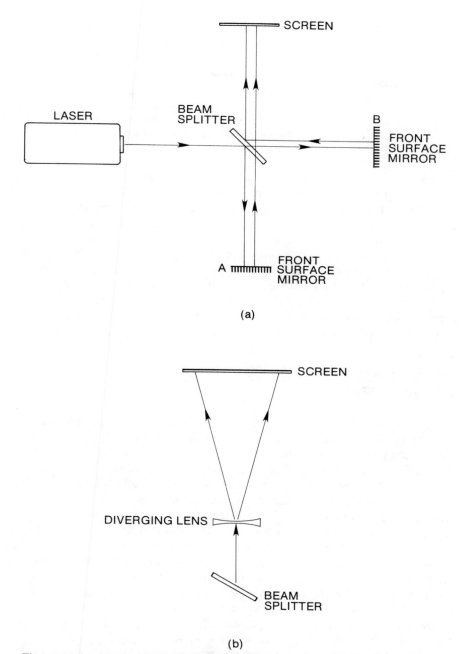

(a)

(b)

Figure 14.1. (a) The Michelson interferometer. (b) After alignment of the undiverged beams, a lens is used to spread out the superimposed beams at the screen.

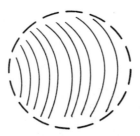

Figure 14.2. *The general appearance of the fringes from a Michelson interferometer as seen on the screen. The curvature of the arcs and their thicknesses depend on detailed adjustments in the system.*

To test for *sensitivity* of the system, try stomping on the floor. Try blowing on the table. Try breathing warm air on one of the components.

The Michelson interferometer, simple though it is, can reveal to you movements on the order of less than one wavelength of the laser light, or less than one-half of a millionth of a meter!

It is a good idea to determine the coherence length of your laser. To do this, gradually move one of the mirrors farther away from the beam splitter. As the difference between the path lengths of the two beams increases, the fringes will wash out. If you continue, the fringes will reappear. The path length difference under the latter condition is one-half the coherence length. The importance of the coherence length is that it limits the depth of a scene that you can holograph.

THE SIMPLEST ONE-BEAM TRANSMISSION HOLOGRAM

This method is discussed in a general way in Chapter 8. It is an excellent first experiment. The configuration is simple and allows you to get experience with manipulating the apparatus, with exposure times, and with the effects of changing details in the geometry, all under the most sure-fire conditions.

Coins are suggested as good objects for your first effort. They are highly reflective, flat, and have surface details that make them interesting to study when viewing the reconstructed image. Place them on a table a few inches from the film holder. Working in a dark room, adjust the positions of the apparatus and the divergence of the beam. About one-fourth of the diverged beam should strike the table in the region of the coins and the other three-fourths should strike the region of the film holder. This can be checked by eye. A piece of paper placed in the film holder is invaluable in helping judge the light distribution where the film will be. Next get a visual idea of what you are going to get in the eventual hologram. To do this, remove the paper and look onto the scene through the film holder. What you see is what you will get. You may want to move the coins around and perhaps tilt some of them to get a bright scene with some degree of vertical depth.

When you are satisfied, put the shutter in place to block the laser beam.

Then look through the film holder again, this time to make sure there is no unwanted light there, as for example due to previously unsuspected reflections from something on the table or elsewhere in the room. If you detect any light block it off.

The next step is to put the film into the film holder. Working in a dark room, or—better—with a dim, green, safety light so you can see what you are doing, put a piece of film between the glass plates of the holder, and put on the clips that hold the sandwich together.[1] Then squeeze to get rid of air trapped in the assembly. Air leaking out slowly during exposure of the film will prevent making a hologram. Squeezing for half a minute or so should suffice.

Now put the film holder in place for exposure with the sensitive side of the emulsion toward the light source. A simple way to find out which side is sensitive is to feel the film with wet fingers before putting it between the glass plates. One side will feel more slippery than the other. The slippery side is the sensitive side.

With the film holder in place, wait long enough to be sure all vibrations you caused have died out. The safe time interval is known from the results of your interferometer test. Then you can make the exposure. First grasp the shutter and lift it so that it no longer makes contact with the table but so it still blocks the laser beam. Wait for the vibrations this operation may have caused to damp out. Then raise the shutter fully, time the period of exposure by counting seconds as you judge them or by using a watch, and replace the shutter in its original position.

How long should the exposure be? This depends on the power of your laser, the sensitivity of the film, and your particular set up, and so no universal rule can be given. With a 1 milliwatt laser something like five seconds is likely to give you an acceptable—even a thrilling—hologram, and then do a little experimentation to optimize the time. With a 5 milliwatt laser, start with one second. The experimentation will give you experience that will help you in making judgments about exposure times with other holographic configurations. Even if you have a light meter, you must do such experimentation to learn how to interpret the meter readings you get.

Having made one hologram, you can try a shorter and a longer time comparing the results. There is another procedure that is easy and will give you more information all in one step. With the film in place and ready for exposure, install an opaque screen to block all but a vertical strip at one edge of the film. Expose for some time interval such as one second. Block the laser beam, move the screen to expose the next vertical strip, and expose again for a second. Repeat with a new strip and a time of a second, and two or three times more. After the film is processed, you will be able to see the effects of various times of exposure.

---

[1]If your film is in the form of a long roll, you must first cut a piece to the size you want. It is a good idea to avoid this delay by having previously cut some pieces, storing them in a light-tight container.

Besides judging the effect of various exposure times by examining the reconstructed images with the holograms, much can be told simply by looking at the developed film in room light. If there is a darkening to a degree comparable with what you see in ordinary photographic negatives, the film is acceptably exposed. With overexposure there is too much blackness, and with underexposure there is only a light gray shade. Notice that these remarks apply only if the film is not bleached during the processing, since bleaching results in a transparent film.

As soon as the film has been processed and dried (Chapter 13), you can view the reconstructed image. Put the film back between the glass plates to keep it flat, replace the holder in its position, remove the coins or other objects, and look through the film. After perhaps some head movement to find the best viewing direction, you can admire your handiwork. We have yet to meet a person who does not respond strongly to his or her first successful hologram. By all means, also look at the real image. With this method, the real image is likely to be striking.

TROUBLESHOOTING

The one-beam-bounce-off-the-table method is likely to give you a hologram on your first try, but it is possible that something will go wrong. As an aid in troubleshooting we offer the checklist that follows. (In order to make this helpful in relation to all of the methods described in this chapter, possible troubles that do not apply to the simple one-beam method are included.)

1. If you do not have a hologram but the emulsion has been exposed, something probably moved during the exposure. If you see some red along the bottom of the film but not elsewhere, you can assume that the base of the film holder held steady while the upper parts vibrated. If your Michelson test showed your table to be resistant to vibrations, examine all of the components and their mounts. Is it possible that one sagged, as can happen if a tube is not deep enough in sand? If you have the laser propped up on something, such as a wood block, do you need shims to ensure against shaking? Are you sure the film-holder plates were squeezed enough before the exposure? Did you wait long enough after first lifting the shutter from the table?

2. Is the film fresh and unfogged and are the chemicals active? Try developing a small unexposed piece of film without bleaching. It should come out clear. Try developing a small piece after exposing it to light. It should come out dark.

3. How long was your laser on before making the exposure? Lasers go through a warm-up period, and for a half hour or more may not be stable enough for holography.

4. The intensity of the object beam may be too low relative to the intensity of the reference beam. With the simple one-beam transmission hologram method and with the simple reflection hologram method this will not be such a critical matter as to prevent getting a hologram. In any event, these methods do not allow much opportunity for adjusting beam ratios.

With other configurations you should compare the beam intensities before exposing the film. The simplest way is to use a card inserted into the reference beam at the location of the film holder with the object beam blocked off, and then to repeat for the object beam with the reference beam blocked off. Because the beam ratio need be only approximately equal to the optimum, it can be adequately judged by eye.

5. Cosmetic defects such as blotches in the reconstructed image are most likely due to fingerprints or other smudges on optical components, to water spots on the film, or to noise in the laser beam itself. Lenses, beam splitters, and mirrors must be scrupulously clean. Wash with alcohol or lukewarm detergent solution, and swab lightly. Water spots on the film can be avoided by using a squeegee or by using PhotoFlo or the like at the end of the processing. The only cure for a noisy laser beam is to use a spatial filter. Remember that the beam must not hit the edges of a lens or other component, because this gives rise to diffraction effects that spoil the quality of the hologram.

6. You may see a pattern called *Newton's rings*. This is set of very numerous closely spaced concentric circles. It can be beautiful, but it is an extraneous effect. It occurs when light that has passed through the emulsion reflects back to it and interferes with the light falling directly on the emulsion. This is how a reflection hologram is made, but in this case the "object" is some flat surface, which accounts for the circular pattern. The reflecting surface can be the dye backing on AH (antihilation) film or a glass surface. If you get the rings with NAH film, you can try a black light-absorbing material to prevent the reflections. Usually holograms are made with off-axis arrangements. Then the rings are not likely to be seen superimposed on the image, and so are not disturbing. In our opinion it is worthwhile to try deliberately to make a good set of Newton's rings to keep on hand to show to visitors, who will be impressed.

ANOTHER ONE-BEAM TRANSMISSION HOLOGRAM METHOD

Figures 1.4, 1.5, and 8.6 show an overhead view of a one-beam configuration in which the reference beam is reflected from a front surface mirror before it goes on to the plate. One advantage in this is that the objects in the scene need not be flat. By using erect objects, such as chessmen or figurines, more interesting scenes can be devised. Another advantage is that severe backlighting is avoided. Still another is that it gives some freedom in adjusting the beam ratio, which should be about 4:1.

The beam ratio can be judged adequately by eye with the help of a white card. Put the card where the film will be later (when you are ready to make the hologram), block off the reference beam and the object beam one at a time with a piece of cardboard or the like, and compare the intensities you see on the card.

With this method (and more elaborate transmission hologram methods), the objects in the scene appear against a dark, empty background when the image is reconstructed. If you want, you can provide a more interesting backdrop by using wood, cloth, a book, or the like.

Finally, there is a caveat. With this method and with methods to be described later, there is danger that some of the diverged laser beam might reach the plate without first reflecting from the object or from the mirror. This can be detected by removing (or blocking) the mirror and the objects, putting a card at the location of the film holder, and looking for light falling on it. Such unwanted light can be screened off by a strategic positioning of a beam-blocker.

REFLECTION HOLOGRAMS

The simple configuration of Figure 8.10 for making a reflection hologram is well worth trying. It is simple and can give you a good white-light viewable hologram.

The intensity of the reference (direct) and object (reflected) beams should be nearly equal. Inevitably the reference beam will be brighter, but if the object is bright red or white in color, and very near the film, the beam ratio will be nearly 1:1. Figure 8.11 offers more control. Moving the lens can reduce the intensity of the reference beam, or a filter can be used, or both. Judging the equality of the beam intensities by eye, with a card inserted into the beams, can be done with accurate results.

With a reflection hologram, light must strike the emulsion from both of its sides. Therefore, a nonantihalation film should be used. (Antihalation film has an opaque layer on one side.)

The exposure time can be made one or two seconds long with good results, if 8E75 film and a one milliwat laser are used, and if the laser beam is spread out to a circle with diameter of about three inches. Under other conditions a little judicious guessing will give you a good approximation to a suitable time. You can get closer to optimization with a little experimentation.

A 360-DEGREE TRANSMISSION HOLOGRAM

Cylindrical holograms of 360 degrees give images that can be viewed all the way around them. Figure 8.7 shows two configurations. Either will give you a hologram that is likely to please you because of its brilliance and clarity, as well as because of the see-around feature.

If the arrangement that has the laser in a vertical orientation is used, it is important that the massive laser be mounted on a stable structure. A single rod (such as a laboratory stand) will shake. Perhaps the most convenient way is to put the laser in a horizontal position on top of a structure made with concrete blocks, bricks, or wood. A front surface mirror is then used to divert the horizontal laser beam into a vertically downward direction. The completely horizontal configuration in part (b) of the figure is simpler, but the object will have to be held firmly to the center of the base of the cylinder, as with epoxy cement.

The cylinder can be any transparent glass or plastic jar, or even a non-reflective opaque one, four or five inches in diameter and about three inches high. (One comes in the Metrologic kit.)

Before installing film into the cylinder, some attention must be given to the beam intensity ratio. As often with transmission holograms, a ratio of 4:1 is desirable. Because the reference beam will be much more intense than the object beam if a dull object is used, a highly reflective red or white object should be chosen. Then move the lens closer to and farther from the cylinder, while you judge the lighting conditions at the cylinder by eye. In doing this have a piece of white paper in the cylinder to simulate the film. This sounds unreliable, but in practice it is highly probable to give you an impressive hologram.

When you are ready to make the hologram, cut the film to size, working in the dark or under a dim green safelight. Put the film in the cylinder with the sensitive side inward, using small pieces of tape to hold it. There should be no overlap or gap if you want a continuous look-around hologram. The film should be in good contact with the wall, and it should stand in place for perhaps five to ten minutes to be sure it is in equilibrium during the exposure. The exposure time is likely to be best at about three or four seconds with a one milliwatt laser, or judiciously more or less with other lasers.

To view the hologram after processing the film, put it back into the cylinder if this is transparent, or simply form the film into a cylinder and hold it together with paper clips or tape. Direct the laser beam down into it with the arrangement that was used in making the hologram, minus the object. Look from outside, through the film, and toward the axis of the cylinder.

An interesting thing to try is to flatten out the film and use it as if it were an ordinary transmission hologram, looking through it back toward the light source. If you hold the film in your hands, curving it as you move the film from side to side in the light beam, you will see interesting distorted images of the screen. The effects are far better seen than read about.

TWO-BEAM TRANSMISSION HOLOGRAM

A two-beam configuration gives the experimenter much more ability to adjust the intensities of light beams, their path lengths, the angles of illumination, and purely aesthetic aspects of the scene. This desirable freedom comes accompanied by a step up in the amount of care that must be taken in setting up the apparatus, but learning to do this is not difficult. Making a two-beam hologram is much easier than it sounds, as is true of many activities which we know to be very simple.

The configuration in Figure 8.8 offers great flexibility because it uses two lenses, one in each of the beams. A good way to begin is to set up the apparatus roughly as you think it will be used but with the lenses not installed. Block off the unwanted beams from the beam splitter and center the remaining beams on the mirrors and on the film holder. You should be able to see where the beams hit the components directly, or you can use a piece of paper to guide you. It is an excellent idea to use smoke at this stage. Smoke (or chalk dust) makes the beams distinctly visible and produces a pretty display.

Next there are two refinements to be made. One is to equalize the beam paths, and the other is to get a large angle—but not too large an angle—between the reference and object beams at the film holder. Too great a difference in the beam path lengths may prevent getting a hologram or cause a dim one. A large angle is needed to capture fine details in the hologram, but with too large an angle the spatial frequency of the fringes may be too great for the resolution of the emulsion. An angle of about 45 degrees would be suitable for a first try.

The next step is to spread out the light beams. Install the lenses, making sure that the beams do not hit the edges of the lenses. The object should be covered with light, and both the object and reference beams should cover all or most of the area where the film will be placed. A piece of paper simulating the film helps very much in judging the illumination of the film holder. The use of smoke to make the beams visible along their lengths can also help.

At this stage attention must be given to the beam ratio. A ratio of 4:1 is about right, though the true optimum ratio depends on details of the configuration. Put some paper in the film holder, block off the object beam and judge the intensity of the reference beam on the paper. Then block off the reference beam and judge the intensity of the object beam on the paper. If the object beam is *much* less intense than the reference beam (the most likely condition at first) try moving one or the other of the lenses. Moving a lens changes the intensity of the beam at the plate. Judging the beam ratio by eye can be done with sufficient accuracy, although no doubt this does not seem likely to the inexperienced. Also, as you can learn by a little experimenting, the results do not depend crucially on the exact ratio.

Now for the last preparatory steps. Measure the path lengths of the beams again, this time measuring from the lenses, and make whatever adjustments are needed to equalize them. Then look through the film holder to get a final impression of what the holographic image will look like. Remember that what you see is what you will get. Block off the reference and object beams and look through the film holder for stray light that might be arriving there. Block off any that is found.

Finally, decide on the exposure time you will use, install the film, and shoot. The exposure time needed will be greater than with a one-beam method. Probably six or seven seconds with a one milliwatt laser will be nearly right, but you may have to try two or three different times to get the best time. You might consider using the strip-by-strip method suggested earlier. Even if you have a suitable (sensitive) light meter for use, you will have to calibrate it by trial-and-error.

Once it has been processed and dried, the hologram can be viewed. The best way is to put it into a laser beam that is spread out with a diverging lens. The slide projector or other intense white-light source can be used if a color filter is installed to give approximately monochromatic light. In order to see the image, you will have to try different orientations in the beam. This is done most easily by holding the film in your hand.

Examine the real image as well as the virtual image. Try turning the film front-to-back so you see both images with this reversed orientation. Look at the images with a white-light source to study the color smears that result.

When you get a good hologram, it is wise to protect it against handling. Our method is to put the film between two pieces of thin glass cut to size and to tape the edges. We have a hologram that has been protected in this way for many years, with no signs of deterioration in the film or in the tape.

GOING FURTHER

The methods we have discussed here can occupy an inquisitive experimenter for some time, improving his or her technique and trying variations. While we have not dealt specifically with the two-beam reflection method or with more elaborate transmission methods, these are feasible with the experience gained with the simpler methods.

Let us suggest some other possibilities open to you from the exceedingly rich field of holography. Make a two-channel hologram with any of the simple methods we have discussed. (Expose the film, shield it from light, change the scene, rotate the film, and shoot again.) Try making a projection or "standout" hologram. Interferometric holograms made by the double-exposure method described in Chapter 11 are simple and can prove to be quite captivating. Try copying holograms. If you are already interested in photography, you may want to try varieties of film. Polaroid (55P/N is a good choice) can be used, though the resolution is not high. Or you may want to try other developing and bleaching processes.

One of the pleasant aspects of holography is that it lends itself to working with others. You may find other enthusiasts. You can give friends and relatives a good time by guiding them through the making of a simple one-beam hologram. There are possibilities for science fair projects for youngsters with whom you can share your lab and experience. Showing your laser and how interestingly the beam can be split and reflected, especially with some smoke in the room to make the whole display breathtaking to the uninitiated, provides a good show.

When you show holograms, you must anticipate the inevitable question, "How do they work?" As we said in the preface of this book, our purpose has been to explain this clearly and simply. It is our sincere hope that we have succeeded.

Appendix I:
Laser safety

Some people are uneasy about having lasers in the vicinity, being perhaps under the impression that all lasers are death weapons. Such fear is grossly exaggerated, but on the other hand the large numbers of people who seem unaware of possible hazards are also sadly uninformed.

What are the facts? First of all, we speak here only of low-power lasers, with output power not exceeding 5 milliwatts. This includes the helium-neon lasers that are readily available on the open market for use in schools for demonstrations and for holography and optical experiments. It does not include pulsed research lasers that can pack millions of watts of power into brief pulses, and that can vaporize pieces of metal.

Low-power lasers present no hazard to humans in any way *except to the eyes*. If the eye should be directed straight into the laser beam (or into a reflected beam) there *may* be irreparable damage to the retina. It is the same as the danger inherent in looking directly into the sun.

Millions of low-power lasers are in widespread use. Workers in construction crews safely use them in leveling ceilings and other structures, in laying long straight stretches of pipe and the like. In countless schools and colleges lasers are used routinely as laboratory instruments.

Laser safety is the concern of the United States Bureau of Radiological Health. This agency makes the following classifications: Class I lasers are usually 1/2 milliwatt or less. Class II lasers are one-half to one milliwatt. Class IIIa lasers are one to two milliwatts. Class IIIb lasers are two to fifty milliwatts or more. Class I lasers do not emit any hazardous light. Class II lasers cannot cause damage to the retina unless the recipient stares into the beam for a long period of time. Class IIIa lasers can cause damage if stared into, or if the beam is concentrated and then brought to the eye, as by use of a lens, a telescope, or binoculars. Class IIIb lasers can produce accidental injury if viewed directly.

A factor to be considered is that the exposure of your eyes to intense light is so unpleasant that a quick reaction forces you to turn your eyes away. An accidental long exposure can hardly be expected to happen. Even so, on the occasions of solar eclipses, the newspapers sometimes report cases of eye damage to persons who *did* look too long at the source of light without adequate intensity-reducing filters. The message of the government agency is that with sufficient laser power, a potential hazard may exist, but it is avoidable if you use common sense.

We assume that lasers used for holography by readers of this book will be at most two miliwatts in power. For users of such lasers the best advice available is that no harm will come from an accidental brief glance toward the laser, as for example when you are trying to pick up an image while reconstructing a hologram, but that at the same time the golden rule is: *Do not look directly into the laser beam even fleetingly.* As a corollary, do not wave a laser around or aim the beam directly or by means or mirrors or otherwise, so that even accidentally anyone present might receive the beam in the eyes. Hazard or not, this is a rule to be observed strictly.

So far we have been speaking of direct, undiverged laser beams. If the beam from a low-power laser has been spread out, as it almost always is in holography, there is a great reduction in the amount of light energy the eye can pick up. For example, if the beam from a one milliwatt laser is spread out to a diameter of six inches or more, there is no hazard in taking the light into your eyes. After all, if this were not so, viewing holographic images that are reconstructed with lasers would be interdicted.

Appendix II: Some sources of materials and information

Lasers

Edmund Scientific
101 E. Gloucester Pike
Barrington, New Jersey 08007

Metrologic Instruments, Inc.
143 Harding Avenue
P.O. Box 307
Bellmawr, New Jersey 08031

Spectra-Physics, Inc.
Sales and Service Center
1250 West Middlefield Road
Mountain View, California 94042

Holographic Kits

Metrologic Instruments, Inc.
143 Harding Avenue
P.O. Box 307
Bellmawr, New Jersey 08031

Discrete Optical Components

Ealing Corporation
Optics Division
2225 Massachusetts Avenue
Cambridge, Massachusetts 02104

Edmund Scientific Corporation
101 E. Gloucester Pike
Barrington, New Jersey 08007

Metrologic Instruments, Inc.
143 Harding Avenue
P.O. Box 307
Bellmawr, New Jersey 08031

Holographic Film and Plates

Agfa-Gavaert
275 North Street
Teterboro, New Jersey 07608

INTEGRAF
P.O. Box 586
Lake Forest, Illinois 60045

Holograms

INTEGRAF
P.O. Box 586
Lake Forest, Illinois 60045

Multiplex Company
454 Shotwell Street
San Francisco, California 94110

Museums of Holography,
   such as:

Museum of Holography
11 Mercer Street
New York, New York 10013

Laboratory Manuals That Go Beyond This Book

Holography Handbook: Making Holograms the Easy Way (1982)
Fred Unterseher, Jeannene Hansen, Bob Schlesinger
Ross Books, P.O. Box 4340
Berkeley, California 94704

A Study Guide on Holography
By Tung H. Jeong
INTEGRAF, P.O. Box 586
Lake Forest, Illinois 60045

Holography Using a Helium-Neon Laser (1979)
By Tung H. Jeong and Francis E. Lodge
Metrologic Instruments, Inc.
143 Harding Avenue
P.O. Box 307
Bellmawr, New Jersey 08031

Experiments Using a Helium-Neon Laser (1981)
(Experiments in laser optics, but not holography)
Herbert H. Gottlieb
Metrologic Instruments, Inc.
143 Harding Avenue
P.O. Box 307
Bellmawr, New Jersey 08031

Other Books

A readable book on optics:
> *Introduction to Light: The Physics of Light, Vision, and Color* (1983)
> Gary Waldman
> Prentice-Hall, Inc.
> Englewood Cliffs, New Jersey 07632

A popular-level book on holography:
> *Understanding Holography* (1978)
> Michael Wenyon
> Arco Publishing Company, Inc.
> 219 Park Avenue South
> New York, New York 10003

A popular-level book on holography and lasers:
> *Lasers and Holography* (1981)
> Winston E. Kock
> Dover Publications, Inc.
> 180 Varick Street
> New York, New York 10014

A thorough and mostly nonmathematical book on holography:
> *Holography and Its Application* (1971)
> Yu. I. Ostrosky
> Mir Publishers, Moscow USSR
> (Available from:
> Academic Press
> 111 Fifth Avenue
> New York, New York 10003)

A short but thorough and technical book on holography:
> *Holography* (1974)
> Maurice Francon, translated by G. M. Spruch
> Academic Press
> 111 Fifth Avenue
> New York, New York 10003

A technical and mathematical book *par excellence* on holography and related topics:
> *Optical Information Processing and Holography* (1974)
> W. Thomas Cathey
> John Wiley & Sons
> 605 Third Avenue
> New York, New York 10158

Index